DALE EARNHARDT JR.

Out of the Shadow of Greatness

MIKE HEMBREE

SPORTS PUBLISHING

Sports Publishing books may be purchased in bulk at special discounts for sales promotion, corporate gifts, fund-raising, or educational purposes. Special editions can also be created to specifications. For details, contact the Special Sales Department, Sports Publishing, 307 West 36th Street, 11th Floor, New York, NY 10018 or sportspubbooks@skyhorsepublishing.com.

Sports Publishing® is a registered trademark of Skyhorse Publishing, Inc.®, a Delaware corporation.

Visit our website at www.sportspubbooks.com

10 9 8 7 6 5 4 3 2 1

Library of Congress Cataloging-in-Publication Data is available on file.

ISBN: 978-1-61321-352-0

Printed in China

For Dad.

CONTENTS

CHAPTER 1

A LEGACY of SPEED

"I never told anybody that I was going to be as good as my dad. I've read that a lot, and I appreciate the comparisons, but that has a backlash when you don't run well every week. I just want to drive race cars and make a living doing it." —Dale Earnhardt Jr.

Photo: David Griffin Photography

The starting point in the racing life of Dale Earnhardt Jr. can be found along Sedan Avenue in Kannapolis, N.C., a small, nondescript textile mill town whose stature was elevated in the second half of the 20th century by the exploits of one racing family.

Martha Earnhardt resides there, at the corner of Sedan and Court. The widow of Ralph Earnhardt, the mother of Dale Earnhardt, the grandmother of Dale Earnhardt Jr., she lives at a Grand Central Station of stock car racing, a birthplace of things fast and furious, a historic landmark along the road to motorsports glory.

At the rear of the Earnhardt home, in a small cinder-block garage, Ralph Earnhardt began the great journey that would lift his family to international auto

Photo: Harold Hinson Photography

racing fame, make its famous son a millionaire many times over, and open the door to a third generation of greatness.

This is the heritage that rides with Dale Earnhardt Jr. at every track, on every lap, into every winner's circle.

Ralph Earnhardt loved racing so much that he became determined to make it a full-time occupation in the 1950s, and he succeeded. Most drivers of his era worked "real" jobs and raced on the side, trying to move dollars around to keep both sides of their lives going. Earnhardt put all his time into the garage behind his house and into his race car, and the results showed. He became a king on southeastern short tracks, and he won NASCAR's Late Model Sportsman championship

in 1956. Ned Jarrett, a two-time Winston Cup champion, described Earnhardt as "absolutely the toughest race driver I ever ran against."

Dale Earnhardt Jr., who would blaze a significant trail of his own in the family's chosen lifestyle, was born a year after his grandfather died. A heart attack killed Ralph Earnhardt in September 1973 at the age of 45. He was still racing, still working on cars in that backyard garage.

His grandson thus races with only the stories of his grandfather and of the big heart that gave out too soon, stories of the man they called "Ironheart."

"I wish I had gotten to know him," said Dale Jr. of his grandfather. "He seemed like a really, really great man. I've heard a lot of compliments about him. It's a shame I never got to know him. I wish I'd had that opportunity—to see him, to watch him race."

Dale Jr. during practice on 10/2/2009 for the the Price Chopper 400 at the Kansas Speedway. *Photo: David Griffin Photography*

As the late Buck Baker, another NASCAR champion, put it, "In those days, you just mashed the pedal to the floor and went." That was the way Ralph Earnhardt raced, and he passed the method along to his son, Dale Sr., who traveled the Carolina racing circuit with his father, watched him tinker in the Kannapolis garage and waited impatiently for his chance to hit the track.

Dale Sr. grew up with only one thing dominating his thought process—racing. He quit school in the ninth grade, convinced he could make enough money to survive while also keeping a small racing operation afloat on Carolina short tracks. He had seen his father do it; in his thinking, there was no reason he couldn't.

"My dad was more independent than I am," Earnhardt remembered in an August 2000 interview.

"**THAT'S MY GOAL. NOT ONLY TO WIN A CHAMPIONSHIP, BUT CHAMPIONSHIPS." —DALE EARNHARDT JR.**

"He was a self-made guy who worked hard for what he got. He was also a control guy. He didn't buy it if he couldn't afford it. He didn't believe in credit."

A determined Dale Jr. was excited by racing from an early age
Photo: Harold Hinson Photography

Dale Sr. didn't have that luxury. He went into debt to race, leaving his garage virtually every race night knowing that he had to make enough money at the track to pay the next week's bills. He also took jobs as a welder and insulation installer.

"You really grow up in racing around with your dad," Earnhardt said.

"Your dad is your idol, and everything revolves around him. You just want to race. My dad was the focus of my life. I didn't like school. I wanted to be home working on Dad's race car. I wanted to be home working on cleaning up the shop. I'd just as soon be washing wrenches. I followed him around with everything he did."

Dale learned quickly from his father. He could figure out a lot about this sort of racing simply by observing his dad—and his cars. He'd walk into his father's shop the morning after a race and check out the car. If there was more mud near the back of the car

than the front, he knew his father had been in the lead most of the night.

When he started racing, he wanted to be the one who placed the dirt, not the one who ran into its cloud.

Dale Earnhardt's never-give-up, never-give-in attitude enabled him to survive the harsh world of Southern short track racing and eventually launch his Winston Cup career. Two years after his father's death, Earnhardt made his Winston Cup debut in the World 600 at Lowe's Motor Speedway.

Dale Earnhardt Jr. was seven months old when his father, still fighting to pay the bills with three young kids to feed, set sail in NASCAR's big leagues. A quarter-century later, Junior would run his first Winston Cup laps in the same race at the same track, but the circumstances would be remarkably, vividly different.

Senior scratched and clawed his way to the top, and when he finally got there he didn't arrive in style. He hitched a ride in a second-line car owned by journeyman driver Ed Negre, and the day, Earnhardt later admitted, wasn't pretty.

"I was probably the most confused and excited person in the country that day," he said. "I had never run anything over 200 laps on a race track, and here I am going to run 600 miles at Charlotte. That was pretty intimidating, really.

"I worked hard driving Negre's car that day. I'll make no bones about it. One of my last pit stops was for water. I was burned out."

He finished 22nd, completing 355 of the day's 400 laps. It was the first stop on a road that would lead to seven Winston Cup championships and more money than all of Kannapolis could imagine.

"He was just a real energetic kid that lived and breathed auto racing," Negre said. "He wanted to drive real bad. He was one worn-out kid when that race was over."

Flash forward 24 years to another May, and Dale Earnhardt Jr. is ready to make his Winston Cup debut. It's the same racetrack, but everything else has changed.

Senior arrived on the scene in near-obscurity; Junior rolls in as the focus of a news media blitz.

It had started months earlier when the Anheuser-Busch brewing company, through its Budweiser beer brand, had won the bidding for the driving services of young Earnhardt. The deal was for six years; the price tag, some said, was between $50 million and

$60 million. In January of 1999, company officials announced that the so-called "Countdown to E-Day" had begun. Junior would make his first Winston Cup start in the Coca-Cola 600 at Charlotte.

The speedway is only a few miles from Kannapolis, grandfather had raced for many years.

Cameras followed Junior's every move, and the assembled media throng, many times the number that had recorded Dale Sr.'s largely ignored first run in 1975, waited anxiously for the next generation to make its

> "FORGET THE MONEY. I LOVE RUNNING AT TALLADEGA, AND I'D GET OUT THERE AND RUN JUST AS HARD IF THEY PAID ME A DIME TO WIN."
> —DALE EARNHARDT JR.

but, in the big historical view, this was a long way from Sedan Avenue. Ralph Earnhardt had started this circle a half-century earlier. His grandson, next to ride with the great Earnhardt name above his driver window, would take the next big step in grand fashion. He was 24 years old. He sat in car No. 8, the same number his

first lap.

In making the announcement that Anheuser-Busch would sponsor Junior's first six years in the Winston Cup series in Chevrolets fielded by Dale Earnhardt Inc., the giant racing operation Earnhardt Sr. had built near his home in Moorseville, N.C.,

Senior talked about his father and the family's racing connections.

"Hopefully, my father would be proud of the family," he said. "I'm sure he would be proud of what he sees.

"My dad said you can take a driver that will drive that car over his head and wreck that race car every race and bring him back down here and calm him down and pull him back a notch or two and make him a winner. You can't take someone who won't drive that thing in the corner and make him drive it in there. You can't do that.

"We've got one of them guys who will drive that thing to no end. We've just got to harness him up and get him tuned down, and he'll be a winner. That's something my daddy told me when I was a kid, and I've always rememebred it."

The foundation having been solidified many years before, Dale Earnhardt Jr. then slid into his race car for the start of what would be a long and winding road. His grandfather had been one of the best ever on dirt tracks. His father, whose steely glare and endless determination had won dozens of races, had made this moment possible, the runaway nature of his success having birthed a massive, vibrant racing organization.

Now the gentleman of the third generation started his engine.

CHAPTER 2

The GARAGE FLOOR

"[Junior is] a great, aggressive driver as far as I'm concerned. I've watched him mature over the last six months or year or so. He's starting to be aggressive when he needs to and be that much calmer when he don't."
—Bill Elliott

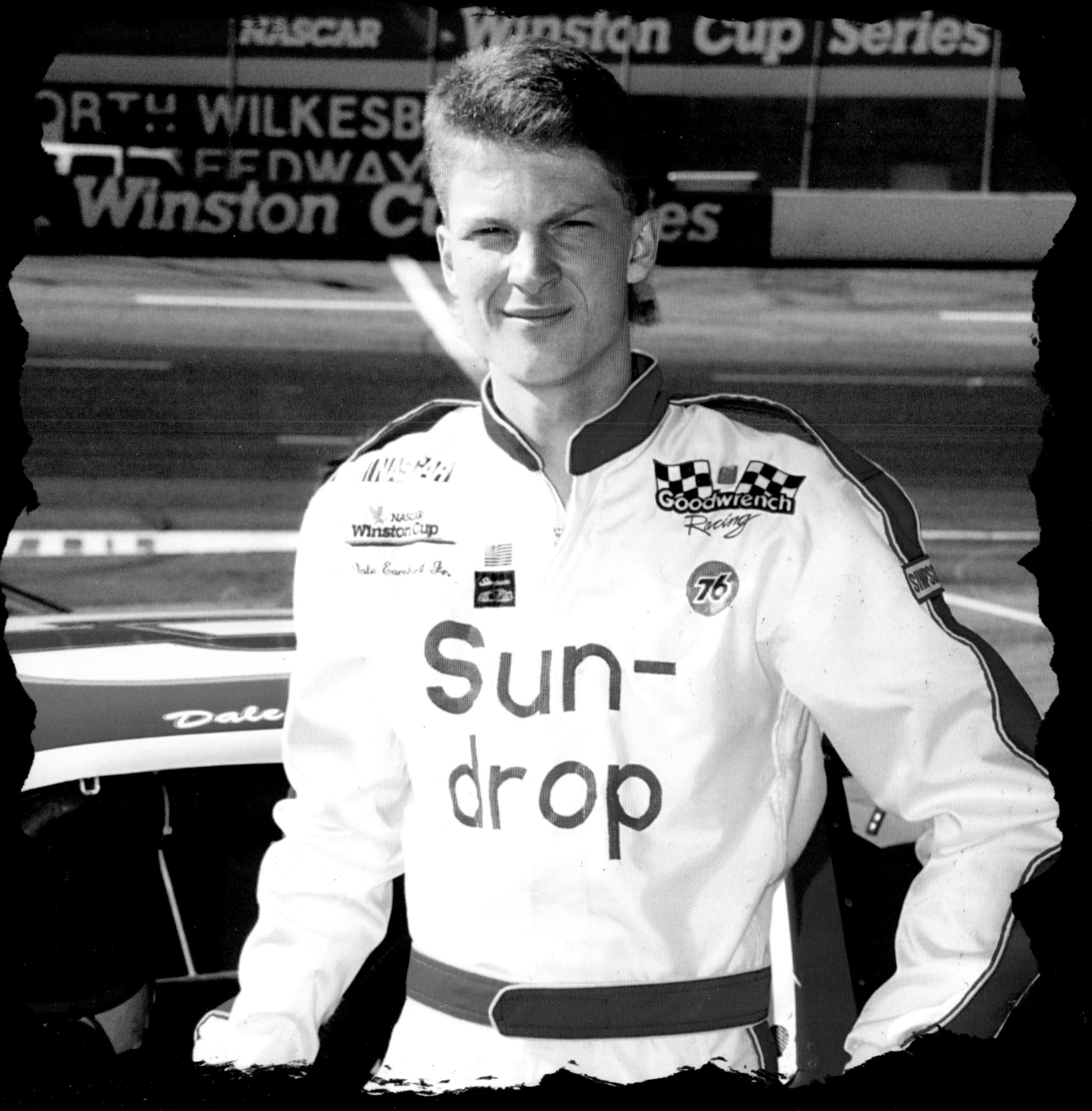

Photo: Harold Hinson Photography

Photo: David Griffin Photography

The road to racing was not paved with gold for Dale Earnhardt Jr.

For more than a few years, there was a myriad of questions about how his life would evolve.

Junior is the son of Dale Earnhardt Sr. and his second wife, Brenda Gee. That marriage also produced Kelley, Junior's older sister. Dale Sr. also had a son,

Kerry, from his first marraige, to Latane Brown.

Dale Sr. and Brenda divorced when Junior was three years old. Junior lived with his mother until he was six, when a fire at their home resulted in Dale Sr. assuming custody of his son.

Junior didn't show significant interest in racing until he was 12. He ran a few races in a go-kart, but his father wasn't too impressed with that form of motorsports and eventually closed that door.

Reaching his junior high school years, Junior found himself in trouble now and then with authority figures, and his father enrolled him at Oak Ridge

Junior sold his racing go-kart for $500. Part of that money — $200 — was used to buy his first race car, a 1978 Chevrolet Monte Carlo from a Mooresville junkyard. Junior was 17. He and Kerry decided to work on and race the car together.

> "THERE IS JUST SORT OF THIS STRANGE PHENOMENON THAT HAS BUILT UP THERE. I LIKEN IT TO BEATLEMANIA."
> —JADE GURSS ON JR.'S ENORMOUS POPULARITY

Military Academy near Greensboro, N.C. with the idea of putting the boy on the straight and narrow path. Kelley joined her brother at the school.

Later, they moved on to Mooresville High School, and Dale received his diploma there in 1992.

Meanwhile, there was some racing to schedule.

"Dale and I were living together in a mobile home on the [family] property then, and we just started talking about going racing," Kerry, later a NASCAR Busch Grand National series regular, remembered. "We just said, 'Let's do it. Let's get a car, and we'll run the Street Stock class.' They had an article in *Stock Car*

Junior and Kerry Earnhardt before a race at Charlotte Motor Speedway in 1998. *Photo: Harold Hinson Photography*

Racing magazine about building Street Stock cars.

"There was a little junkyard down the road. I was always getting parts there. We talked to them and looked around and found a real nice '78 Chevrolet Monte Carlo. It was in perfect shape. I don't know why it was in the junkyard. There weren't any dents on it or anything. The motor even ran. We took it home and pulled it up to the shop and parked it next to the garage [at Dale Sr.'s racing shop].

"Dad came by and saw it and said, 'Don't touch this. I'm going by the farm and I'll be back. This is a nice car. We need to take a look at it.' Dale and I put our heads together and figured out that he was getting ready to take the car away from us. So we started busting out the headlights and windows and pulling the chrome and everything off the car before he could come back and take it away. Of course, we got a

chewing out, but we still had a race car."

The half-brothers worked on the car in the so-called Deerhead Shop (named because it contained some of Senior's hunting trophies) at Dale Earnhardt Inc. They put another $250 into the car for parts, then hauled it to nearby Concord Motorsports Park, a popular short track, and traded weekly rides in the track's Street Stock series.

"I drive it the first week, and then he drove," Kerry said. "We took turns. Dad jumped in and worked on it with us. He welded the roll cage in."

The next year, the boys moved up to Late Models, and they were joined by Kelley, who also drove. It was an adventure, one that didn't always end with rainbows. In one memorable race at Nashville, Tenn., all three Earnhardts crashed in a big pileup on the first lap.

The process of producing the next generation of Earnhardt racers had begun.

And it was done the Earnhardt way.

By the time Dale Jr. reached his teenage years, his father was a multiple Winston Cup champion, had built a successful motorsports operation and possessed a huge fortune, fueled by on-track prosperity, runaway souvenir sales and smart investments. He was a money machine. He had married again (to Teresa Houston), and his life was rolling along in grand splendor.

Senior thus easily could have afforded to hand his son the keys to the kingdom, providing him with great race cars, professional mechanics and every other piece needed for success from day one. But he knew that was not the proper road to take, for he had seen from his own upbringing that struggle and learning were key elements in building a solid foundation. He wanted Dale Jr. to learn from the ground up.

So Junior started on the first floor—the garage floor. He learned how race cars worked, what made them fast, what corrections they needed, how they changed from track to track. His father helped, but he didn't make the going easy.

There was no silver spoon—plastic, if anything, Junior once said.

Kevin Triplett, now a Bristol Motor Speedway executive, worked several years as Dale Sr.'s public relations representative. He was with Earnhardt practically every day during the years when the three Earnhardt children were getting their feet wet on Carolina short tracks.

"His big goal was for them to learn as much as they could about everything—all the way down to how the roll bars went in," Triplett said. "He did a lot of welding when he was younger, and he helped them weld the roll bars in, but he wanted them to be involved so they would know how.

Photo: David Griffin Photography

Photo: David Griffin Photography

"I don't know that I ever heard him say, 'This [racing] is what you're going to do,' but I did hear him say, 'If you're going to do this, then you're going to do this right, and this is what I think is right.'

"He wanted them to learn their way like he and other drivers did. And that's what they did. He never pushed it on them."

Many fans probably had the perception that Junior and his siblings got the best of everything right away. "But it was just the opposite," Triplett said. "I almost feel like I'm violating something talking about it because I felt privileged to see that it was almost the exact opposite of what everybody thought it was. It was neat to see him work with them. I knew how much he respected his dad. I know his dad was one of his heroes. If the way his dad did it was good enough, then I'm sure he felt like it was good enough for his kids."

The old '78 Monte Carlo, then, became the first entry for the next generation. "That was their car," Triplett said. "They ran that horse. They drove it until

Photo: David Griffin Photography

it couldn't hardly go any more."

When Junior moved on to Late Model Stock racing in 1994, his world expanded. He raced at tracks in the Carolinas, Tennessee, and Virginia while establishing himself as a weekly runner at Myrtle Beach, S.C. Veteran mechanic Gary Hargett of Marshville, N.C., assisted Earnhardt Jr. (and housed his car in his shop) in those early runs. In 1995, mechanic Wesley Sherrill joined Junior, and they worked on the car in a corner of Dale Sr.'s shop.

"His dad pretty much put us in the ballgame and said, 'Here it is. Learn to work on it and make it roll,'" Sherrill said.

Dale Earnhardt Jr. at North Wilkesboro, North Carolina, in 1993, at the beginning of his racing career. *Photo: HHP/Harold Hinson*

"A lot of what we did was hit and miss. Dale Jr. had some decent notes from his time with Gary Hargett, and his memory is good. And we talked with a lot of people who had been at the tracks we went to.

"We'd come in on Monday and get the cars cleaned up and talk about what we did wrong with Dale Sr. Then we'd get set up again for Florence [S.C.] and Myrtle Beach. We worked off a budget just like any other race team. We didn't have all the horsepower in the world."

Junior ran 34 Late Model Stock races in 1994, scoring one win and four poles. He was winless in 1995 but rebounded with a pair of victories in '96. Most importantly, he was learning every week.

While in high school, Junior had worked at his father's dealership as an oil changer. He also worked on the family farm, putting in time as cleanup man in the horse barn. He figured out quickly that racing was the way to go.

"I didn't learn enough [during his Late Model years]," Junior said. "I wish I'd paid more attention than I did. Wesley and me were about equally competent. We learned just by having a little common sense and by trial and error."

Randy Earnhardt, Dale Sr.'s brother, served as overseer of the Late Model racing exploits of the three Earnhardt kids. "He was a big supporter of mine," Junior said. "When a lot of people didn't give me the benefit of a doubt, he always did."

Randy handled purchasing for the teams and worked on spreading their schedules beyond the tracks on which they started. "Looking at it back then, I never really thought Dale Jr. was serious enough to handle the whole situation of being a driver with the press and everything else, but he has been a good surprise to me and all of us here at DEI," Earnhardt said.

Photo: David Griffin Photography

"He was real excitable as far as approaching driving a car. That's all he ever focused on. All he cared about was driving that car. He had decided what he wanted to do. He didn't want to work in that dealership. He didn't want to get a job anywhere else. He was just like his dad."

The years banging around short tracks proved valuable. Earnhardt Jr. learned what to do, where to go, which drivers to challenge, when to talk, and when to shut up.

"I think the years he raced his own Late Model are the years when he learned the most," said Tony Eury Sr., Junior's Winston Cup crew chief. "His daddy made him do his own work. He was doing his own adjusting at the racetrack, and you learn a lot that way. When they started going to different tracks, that helped him a lot. You have to make changes, and you can only learn by doing it."

After all those weekend nights making circles at southeastern short tracks, after all that time looking up from the garage floor at the bellies of race cars, Junior was ready for the next level — NASCAR's Busch Grand National series.

Photo: David Griffin Photography

Dale Earnhardt Jr. at Lowe's Motor Speedway in Concord N.C. in 1999. *Photo: HHP/Harold Hinson*

CHAPTER 3

BEATING THE BUSCH

"He's very talented, obviously. I observe him doing things out there that are pretty impressive. He's just learned a lot by watching on TV when he was growing up and being at the races on the weekend, and then he could apply what he learned directly to himself when he would get the car and he could perform."

—Michael Waltrip

Photo: HHP/Harold Hinson

Photo: David Griffin Photography

By the summer of 1996, Dale Earnhardt Jr. had run enough laps in Late Model Stock racing to earn a shot at NASCAR's No. 2 series, Busch Grand National (later known as the Nationwide Series). He hadn't proved that he could be a professional racer and an established star at stock car racing's highest levels, but the promise clearly was there.

Dale's father made the decision to give Junior his first BGN experience at Myrtle Beach Speedway on the coast of South Carolina. It was an obvious but smart and practical move. Junior had cut his racing teeth in the track's weekly racing programs, so he wouldn't be riding in complete darkness in his first chance to challenge the BGN regulars.

The race was June 22, 1996. A popular stop on the BGN circuit in part because it allows fans and competitors a side trip to one of the southeastern coast's most famous beaches, Myrtle Beach Speedway was auto racing central for one special summer day. The son of The Intimidator was in town for much more than a weekly show, and the results would tell much about the speed with which he could continue up the ladder.

Junior's familiarity with the track was a boon in qualifying, and his DEI-supplied Chevrolet was up to the task. His time trial speed was seventh fastest.

The day was a success. Junior ran with the lead group but lost a lap in the pits and finished 14th. It was something to build on.

Dale Earnhardt Jr. at a practice in 2004. *Photo: HHP/Harold Hinson*

The following season, DEI ratcheted up its BGN program for Junior, entering him in eight races as a teammate to Steve Park, who would finish third in the series point standings. In a partial, feel-your-way-around kind of season, Earnhardt Jr. logged a bundle of educational laps and scored his first BGN top 10 — a seventh at Michigan International Speedway.

The 1998 season was Dale Jr.'s first full-time run through the BGN schedule. Although he technically wasn't a rookie since he had run eight races the previous season, he had much to learn and would be running on tracks he hadn't visited. Steve Park had been promoted to a DEI Winston Cup car, so Junior carried the organization's hopes for a BGN championship.

He also carried his father's famous number — 3 — on the side of his car. His dad had driven the same car in the BGN series before devoting his full time to Winston Cup racing.

"I have always felt that I could race at this level,"

Earnhardt Jr. said. "I know I am a rookie, but I've been watching for a long time—picking up stuff along the way. I feel I can get the job done.

"This means a lot to me, to drive for the same team my dad did. I am honored to just have a chance to prove myself. I want people to see that I am my own person."

Earnhardt Jr. would make his move in the best of equipment (built in the sparkling DEI shop) and with the best of crews (longtime Earnhardt associate Tony Eury Sr. was the crew chief). Eury and Earnhardt Sr. grew up together in Kannapolis and raced together on dirt. Eury came to work full-time for Earnhardt in 1986 and helped him build DEI into a powerhouse. It was only natural that Eury be chosen to show Earnhardt Jr. the right roads in the BGN series.

"He's had some real good drivers in that car over the past few years," Junior said of his father's BGN operation. "It makes me feel good that I'm in it. I think I'm ready for the job. I would hate for anybody

to perceive it as something being handed to me. I just look over the past few years at what kind of work we've put into it to get to where we are. I feel like everything's running along to form."

Even then, unproven on the big tracks of NASCAR's high-profile divisions, Junior had a vision of how he wanted to race.

"If I'm behind some fellow, I want him to be thinking he has no idea what's going to happen," he said. "That's the way I want it to be. He's going to be thinking about the wrong thing.

"Dad earned everything he's got with hard work. He's very stern. He knows what you're capable of doing, and he expects that every time."

The season's first race, as always, was a 300-miler at Daytona International Speedway in Florida. It would be Junior's first run at one of motorsports' most famous facilities, a track where his father had shown staggering dominance over the years.

On the Thursday before the Saturday race and Sunday's Daytona 500, the opening event of the Winston Cup season, Earnhardt Sr. won a 125-mile qualifying race for the 500. That afternoon, he visited the press box high above the track to talk about that race and the one to come on Sunday. Below him, Junior and other BGN drivers were practicing for the Saturday season opener for that series.

Earnhardt the Elder watched with obvious pride as his son circled the 2.5-mile track, searching for the best groove for the weekend. It was a track Earnhardt knew better than his backyard. Junior was still searching, feeling his way.

"Look at that," he said. "Imagine if that was your kid in that car down there. Go, June Bug [his nickname for Junior]. Get up there in the draft."

As for specific advice, Senior said he didn't have a lot for his son. "It's really hard to give advice," he said. "It's easier for them to learn by experience. You can teach basics or whatever, but to teach someone the actual hands-on and how it's going to turn out, that's hard to do. You've actually got to touch and feel and

> "AT ALMOST EVERY AUTOGRAPH SESSION WE GET AT LEAST ONE GAL THAT COMES UP AND IS EITHER WEEPING OR SHAKING OR UNABLE TO SPEAK,"
> —JADE GURSS, DALE JR'S PUBLICIST

The Winston winner Dale Earnhardt Jr. at Charlotte Motor Speedway in Concord, N.C. with his father in 2000. *Photo: HHP/Harold Hinson*

experience."

On Saturday, Junior's first full BGN season started in topsy-turvy fashion, literally. On lap 106, Buckshot Jones and Dick Trickle made contact on the backstretch. The impact sent Trickle's Chevrolet into Junior, whose car turned and flipped, climbing into the sky. It landed on its two left wheels before settling upright onto the infield grass.

Earnhardt Jr., whose father had taken similar rides along Daytona's backstretch, wasn't hurt, but he finished a dismal 37th. Earlier in the race, he had roared into his pit too fast and hit his jack man, then

Photo: David Griffin Photography

twisted the drive shaft leaving the pit. It was not the sort of start that leads to championships.

Still only 23 years old, Earnhardt Jr. was also still searching for his place in the world. He was operating on an entirely new level now, racing virtually every week against some of the best drivers in the world, and he was also fitting into a new environment, one of new people, new challenges and different attitudes.

Earnhardt Jr. showed up at one race with a photograph of a girl he had been dating and started showing it around the garage—like a young kid proud of his girlfriend. It was a bit of naivete for Junior, who found himself subjected to merciless kidding from veterans such as Randy LaJoie, a fun-loving BGN regular who is always on the prowl for unsuspecting victims.

Earnhardt Jr. learned quickly, though, and in more ways than one.

In March, he won his first BGN pole, at Bristol, Tenn. Then the next week, the dam broke.

On April 4, in the Coca-Cola 300 at Texas Motor Speedway, Junior charged to the front on the white-flag lap and scored his first BGN victory. His father, working the team radios and talking to his son during the race, smiled from ear to ear.

The Texas crowd went bonkers, and the Dale Earnhardt Jr. bandwagon started rolling.

Before, Junior had the name, the promise and the potential. Now he had the results.

On that April day in Texas, the true significance of his future was underlined. In the minutes after the race, Dale Jr.'s souvenir trailer at the speedway sold out of every product in the inventory. T-shirts, caps and model cars flew off the shelves. One of the first telephone calls from the DEI offices after the Texas victory was to order a larger Junior souvenir rig.

The rest of the season verified Junior's Texas two-step. He won six more times, finished in the top

Photo: David Griffin Photography

five 16 times and sealed the seasonal championship by cranking his engine for the final race of the year at Homestead, Fla. He finished 42nd that day, but it didn't matter. The championship—his first—had been won.

Earnhardt Jr. had not only learned to win, but he also had picked up the fine art of point racing, a talent perfected by his father.

In June of '98, Dale Jr. signed a five-year contract to drive for his father's team with the understanding that he would race again in the Busch series in '99 and probably move up to Winston Cup in 2000.

Later in the year, his Winston Cup future came

into clearer focus. On Sept. 21, it was revealed that Anheuser-Busch and its Budweiser beer brand would be Earnhardt Jr.'s sponsor for a five-race run in Winston Cup in 1999 and for a full five-year program to follow. The deal, negotiated by then-DEI president Don Hawk, was one of the most impressive in the history of the sport, particularly since it involved a driver who had never appeared in a Winston Cup race.

Devoted Earnhardt fans salivated at the thought of father and son racing against each other in Winston Cup machinery. The real deal was still a few months away, but a preview was set up—almost a world away.

Dale Earnhardt congratulates his son for being the first rookie to win the Winston at Lowe's Motor Speedway in 2000. *Photo: Harold Hinson Photography*

In November, Senior and Junior met for the first time on a racetrack — in Japan at the Twin Ring Motegi course. The race was part of a three-year exhibition series NASCAR scheduled to spread its influence to the Pacific Rim. Mike Skinner won the event, but of more interest to many in the crowd (and back home in the States) was the fact that Junior finished sixth, two spots in front of his father.

The plotters and planners at DEI figured Junior needed another full season in Busch racing to hone his talents for Winston Cup, so they prepared another fleet of No. 3 Chevrolets to give him a shot at repeating his championship in the 1999 season, his final full year in the series. Although the year was a little crazy because of the publicity surrounding his debut in the Winston Cup division in May, the team settled into a rhythm that once again carried Little E to championship territory. He won six times on the way to winning the title again.

"I hope he keeps his head screwed on," Dale Sr. said of his son's second championship. "He's got a lot of future ahead of him."

Junior and his father now were fellow travelers at the top of the game.

CHAPTER 4

INTO *The* STORM

"*Dale Jr. is an aggressive driver, and so was his dad. His dad came up in a different era. His dad always wanted to intimidate everybody. He gained that edge just from intimidation. If you're the guy looking in the mirror in front of Earnhardt, you're thinking, 'What's he going to do?' Where, in today's environment, if that gets out of hand, [NASCAR puts] a stop to that. So to me, it's a different era.*" —Bill Elliott

Photo: David Griffin Photography

The publicity building toward Dale Earnhardt Jr.'s debut in the Winston Cup series was unlike anything stock car racing had known. NASCAR, one year after celebrating its 50th anniversary and with its popularity booming, now had a bright young star with regal bloodlines to showcase, and the public relations machine seemed to work in perpetual motion and from all angles.

Never in the sport's history had a driver's first race been so completely catalogued. Although his father had won seven Winston Cup championships and was on everyone's list as one of the greatest racers ever,

Coca-Cola 600 week in May of 1999 belonged to Dale the Younger.

From the closing months of 1998 through the winter and early spring of 1999, Junior answered a

to Junior's because the news media presence in the sport had multiplied dramatically over the two decades and fan interest had exploded.

"Now instead of 50,000 people in the stands,

> "[JUNIOR] DRIVES A RACE CAR EVERY LAP ON THE RACETRACK AS HARD AS IT WILL POSSIBLY GO."
> —LARRY McREYNOLDS

thousand questions about his arrival in Winston Cup. Was he worried? How different would it be? Were people expecting too much? What advice did his father give him? Was he afraid of failing to qualify? Would the longer race make a difference?

Not since Kyle Petty followed his father, Richard, into the sport in 1979 had a namesake been under such scrutiny, and Kyle's arrival couldn't really be compared

you've got 150,000," Kyle Petty said. "Instead of being in six newspapers, you're in *USA Today* and every national paper. You're on [ESPN's] *SportsCenter*. It's that kind of deal. I think the fishbowl got tighter. There's more people looking in the fishbowl than ever before."

Petty, who has some significant experience in the concept of sons following fathers into racing (he

Photo: Harold Hinson Photography

followed his father, who followed his father Lee, and Adam, Petty's son, raced into the fourth generation) said Junior handled the fame his name carries with grace.

"Talking to Junior, I really don't think it affected him," Petty said. "It may be different this year [1999] because he has a little bit more on his shoulders. People are looking at him for different reasons, but last year watching Dale Jr. and the way he handled himself with the press and around other drivers, he was just another driver having a big time. He just happened to be Dale Earnhardt's son."

The ramp up toward May and Charlotte worked on Junior, however. He was racing full-time in the Busch series and also fulfilling appearance commitments for Budweiser, his Winston Cup sponsor. His schedule was scripted almost by the hour, and before he ran his first Winston Cup lap, he was almost overwhelmed by it. In addition to racing and testing, his schedule included 18 appearances for ACDelco (his primary Busch sponsor), 12 for Budweiser, eight for Coca-Cola, and several for a dozen other sponsors. He jokingly suggested starting a self-help group called DABS—Drivers Against Busy Schedules.

In several interviews prior to the Charlotte race, including one that appeared in *Sports Illustrated*, Earnhardt Jr. voiced concern that the hype machine had gotten out of control, that he no longer had time to think about his racing and the preparation for it. Steve Crisp, a longtime DEI employee and the person in charge of arranging Junior's schedules (he carried a big folder labeled "Dale Jr. Stuff"), stepped in to do some damage control, creating relief-producing off days for Earnhardt Jr. between racing, testing and sponsor appearances.

At 24 years old, Junior didn't want to be the victim of burnout before he started a Winston Cup engine for the first time.

"There's been a lot of stuff going on, building up to this," he said. "Now we're to the point that everybody's real excited about it and real anxious to get going."

For him, that was a large understatement.

He fell back on the presence of his father.

"This is pretty special to Dad," he said. "That's something I enjoy about it, knowing that it's important to him. That makes it a lot easier to handle all the pressure and the hype and the buildup to it. He's always been real special in my life and been a big, big part of it. It's very comforting to know he's still a big part of it going into one of the most important days of my life."

No one expected Junior to contend for the win in his first time out, particularly in the season's longest race. But he didn't expect to languish in the rear, either.

"I've watched these races pretty closely in the last five or six years, and you usually get three or four guys who are a lot faster than everybody else, and the rest

of the field has a tough time staying on the lead lap," Earnhardt Jr. said. "If you can stay on the lead lap and have good green-flag pit stops, you can do well. It's our first race, and we're pretty much prepared for it as best we know how."

Then, one final time before the big day, he was

to spread out a little more than that."

Senior's expectations for his son entering the big Charlotte event were not demanding. "I want him to learn," Earnhardt said. "I hope he impresses me, but I just want him to learn. Stay out of trouble and learn something.

"He doesn't remind me of me yet. He ain't that

> ## "DALE JR. HAS SO MUCH TALENT, NATURAL TALENT. HE'S ALWAYS A GUY YOU'VE GOT TO WATCH OUT FOR."
> ## —MARK MARTIN

asked to compare himself with his father for perhaps the thousandth questioner. "We're not that much alike," he said. "We don't like the same kind of music, and there's a lot of other things, too. I like Pearl Jam, alternative rock, Elvis Presley, '70s soul and rock music. Dad, he just listens to country. I think he needs

mean yet. He's done a great job, though. There's a lot of pressure on him. There's a lot to do. We're trying to keep some of the people around him throttled back.

"It's unbelievable the way your kids grow up and the way things change over the years. I'm having fun with it, enjoying seeing him do what he's doing.

"I was 29 when I came into Cup [full-time]. I had already beat up and down the road in the Sportsman division. I spent my money until I was broke. I didn't have nothing. It's a little different for him. Things are there for him—his team, his sponsors, things like that."

Before he cranked his Chevy to start that first race, because of lucrative sponsorship deals and crafty marketing moves by DEI, including the release of die-cast models and assorted souvenirs bearing the Earnhardt image, Junior had earned more money than many also-ran drivers make in a career.

Leading to the race, Earnhardt Jr. was in a Charlotte parade, made a race festival appearance to play drums with a rock band made up of some of his friends and generally was the big man in town for the week. At the track, he was swarmed by fans.

Everywhere, they wanted autographs. A few minutes before the race, veteran driver Ken Schrader, a close friend of Earnhardt Sr., asked Junior to autograph Schrader's driving suit, making a permanent reminder of the big day.

After the race actually started, the Junior furor died down. He had qualified a surprisingly strong eighth—after some fears that he might not make the starting grid, which would have been a large embarrassment—but dropped through the field quickly. Fifty-five miles into the long day, he had dropped out of the top 20.

Junior battled a loose condition (meaning the rear end of his car tended to slide through the turns) most of the race, ultimately finishing 16th. It wasn't magical, but it was more than acceptable for the first time out in such a grueling race and under such immense pressure.

Twenty-four years earlier, Earnhardt Sr. had finished 22nd in his Winston Cup debut in the same race. The numbers were roughly comparable, for anybody who wanted to do that sort of figuring.

Dale Earnhardt Jr. and Jeff Gordon discuss Busch series strategy in 1999. *Photo: Harold Hinson photography*

Junior probably would point out, however, that he didn't have to pit late in the race for a water break in what was his first long-distance event.

Junior ran four more Winston Cup races in what became his second Busch Grand National championship season, finishing 43rd at Loudon, N.H., 24th at Brooklyn, Mich., 10th at Richmond, Va., and 14th at Hampton, Ga.

The five Winston Cup races in 1999 didn't present father and son Earnhardt with a real opportunity to race each other, but the year didn't pass without such a grand moment. And it was pure theater, a signature moment that, considering events that would impact the Earnhardt family in the years to come, would become even more special than it seemed at the time.

The site was Michigan International Speedway. Both Earnhardts were participating in the International Race of Champions, a made-for-television series that matched drivers from various forms of motorsports in

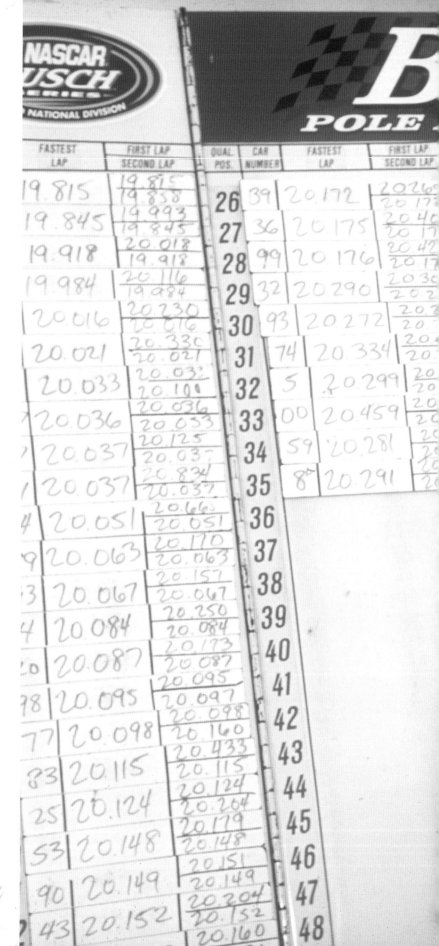

Dale Earnhardt Jr. at Myrtle Beach Speedway in Myrtle Beach, S.C. *Photo: HHP/Harold Hinson*

Photo: David Griffin Photography

identically prepared cars in short sprint races. The June IROC race at Michigan resulted in an all-Earnhardt show. Sen ior was a veteran of many IROC races and a former champion; Junior had been invited to participate because of his Busch Grand National championship the prior season.

The Earnhardts ran one-two—Senior first, Junior second—entering the final lap. Approaching the last turn for the final time, Junior jumped to the high side of the track to attempt to pass the old man. They were side by side for seconds, then they bumped. Junior lost control for a split second, and Senior won the race by about two feet.

They later disagreed about the bumping and who was at fault, but it wasn't a big deal, anyway. There would be many other opportunities for the two to race with the checkered flag in sight—or so it seemed then.

Junior joined his father in victory lane. They laughed and joked about the finish, and Senior later said there was no way he would have "let" his son win the race.

"Wouldn't have been cool," he said.

"I never thought about trying to go by Junior. It's a trust thing that he and I have. I know that if I'm leading the race, he has always stuck with me, and he knows that any time I've been behind him like that late in a race that I've always stuck by him."
—Tony Stewart, 2002 Winston Cup Champion

CHAPTER 5

A ROOKIE And A ROCKET

"He may be Junior, but he's got all the talent in the world, he's got all the talent Senior had. I knew it was me and him [for the win]. I knew it'd be a shootout."

—Tony Stewart

Photo: *Harold Hinson Photography*

The 2000 Winston Cup season dawned with Dale Earnhardt Jr. officially assuming "rookie" status and with the public pressure on him showing no signs of relenting.

The news media followed his every word and move, his fan base seemed to grow geometrically, and his team was ready to make the step up from the Busch series to full-time Winston Cup. It would be a big jump, mechanically, organizationally, and emotionally.

For Junior, who had stepped gingerly into NASCAR racing as a relatively shy, honest kid who wore his famous last name well, the changes were immense. There were questions about how he would handle the dazzle of it all.

"A few years ago I was driving Late Model cars," he said on the eve of his first full Winston Cup season. "Monday through Friday I could do whatever I wanted to do. A lot changed, and a lot changed fast. I'm just trying to adjust to it.

Junior and Senior await driver introductions for the drivers eligible to win the Winston No Bull Million at Richmond International Raceway in 2000. *Photo: Harold Hinson Photography*

Photo: Harold Hinson Photography

"I know when it's time to wear a suit and when it's time to wear jeans and a T-shirt. The biggest problem I have is trying to stay the same in here [his head] and in here [his heart]. I'm trying to have the same values I had even though I'm not backing up my power bill two weeks late like I was two years ago.

"The hardest thing for me is to realize the value of a dollar and the value of a friend and to be able to tell who's good and who's bad. Things like that—things that were easy to do—aren't easy to do any more. That's the hard part of it."

At 25, he was becoming something of a philosopher. Beyond that, though, how would he deal with the smothering pressures of the first season, the heavy burden of his name, and the punishing travel and driving and sponsor demands of the long schedule?

Some wondered, also, if Junior's DEI crew, elevated from the BGN series to the big top, was ready for the chase. There also was the issue of the team's race cars — Chevrolet was debuting a new Monte Carlo model for the new season.

From the Earnhardt inner circle, there was little doubt that he was on course.

"He's very similar to his dad in that he'll do whatever he's got to do," said Steve Crisp, DEI's watchdog over Junior's early career. "He won't settle for second place. He'll drive the wheels off the car, and if he bends one every now and then, he's going to do it. He's extremely competitive like his dad, and extremely focused like his dad."

Tony Stewart, an open-wheel champion who had detoured from that form of racing to join NASCAR, had established a big target for Earnhardt Jr. the previous season. Running as a rookie, Stewart won three times, becoming the only first-year driver to do so. He also finished a surprising fourth in the point standings.

"I don't see me being as hot out of the box as he was," Earnhardt Jr. said of Stewart. "A good year would be a top 20 in the points, running in the top 15 consistently and kind of stepping that up as we go.

"We're going into our first full season. I'll be a rookie. I'm driving for my father, a seven-time Winston Cup champion. So there's a lot of prestige there. There's a lot of good expectations there. We're capable of putting together a good season. I'm going to enjoy it and have fun with it regardless of what happens."

Junior had no illusions. He knew he had a world of adjustments to make, not the least of which was the move into the bigger, more powerful cars in the

Celebrating the first rookie to win the All-Star event known as the Winston
at Lowe's Motor Speedway in 2000. *Photo: Harold Hinson Photography*

TIONAL GUARD

amp ENERGY

GUARD

Mtn Dew

IMPACT

Winston Cup series. Additionally, he would be racing on tracks he had never seen.

"There are certain ways you drive a Winston Cup car that's different from a Busch car," he said. "I went to Atlanta [the previous season], and it took me about half the race to figure out a faster way around the track in a Winston Cup car. I was driving it like I drove my Busch car, and there's a faster way. It just took me a while to figure out by watching those other guys.

"We're going to have to go to all these tracks I haven't been to, so I've got to learn that. I've got to become a little better about the feel of the car and get the car adjusted to make it faster. We can't rely on the setups and the changes we've had with the Busch car.

"It's all here. I've just got to improve and get on the racetrack and go a little bit quicker. These other teams are pulling on the racetrack and clicking off [fast] qualifying laps right off the bat. It takes me five to 10 laps to get accustomed to the track and up to speed. I

need to work on that.

"You find out by doing. It's on-track experience. That's the only way. You can talk and watch all you can until you get out there and actually do it—that's when you'll find out."

The big thing, Dale Jr. figured, was to be in position to make noise at the finish. It had been the key to his father's long run of success. "I've been watching for about 15 years as hard as I can watch, and he just knows," Junior said. "It's all about putting yourself in position to win. That's what he does. He knows where he wants to be and when."

Although it was a secondary issue, the season also held the promise of an interesting battle for NASCAR's Rookie of the Year award. Earnhardt Jr.'s main competition would come from one of his best friends, Wisconsin native Matt Kenseth, a smart young driver who had landed a top ride with the Roush Racing team. Earnhardt Jr. and

Richmond winner Dale
Earnhardt Jr. in 2000.
Photo: HHP/Harold Hinson

Kenseth had staged some memorable battles while moving through the Busch series together. Like Dale Jr., Kenseth had prepped for 2000 by driving in five Winston Cup races in 1999, scoring a high finish of fourth.

Junior's fan base wasn't talking about Kenseth, however. They were anxious for their man to get to the Big Show and lock horns with the leader of the pack, Jeff Gordon. Three times a Winston Cup champion entering the 2000 season, Gordon had brought a fresh new look to NASCAR racing. An open-wheel ace who decided to switch to stock cars, he had stormed into NASCAR and started fires almost immediately, winning races and championships with flair and pizzazz. But he was everything the typical Southern NASCAR racer wasn't (he didn't wear cowboy boots or hat, didn't hunt, didn't fish), and that made him an easy target for the millions of fans who had followed Earnhardt Sr. for years and who now were looking to Junior as a new challenger

for Gordon.

Junior appreciated their intentions but didn't plan to begin shooting for Gordon on his first lap on the track. "When people mention they want me to get up there and battle Gordon and take care of Gordon, it makes me think if I was that competitive in one single event this year, I'd be happy," he said. "This stuff about Gordon is kind of strange—just that people are thinking about that so early."

The Earnhardt Nation clearly expected Junior to win races his first season. Other people weren't so sure. There was much to learn. The competition was fierce. As a rookie, he would have to prove himself to veteran drivers. All this—and with a camera stuck in his face much of the time.

Yet he did it.

The season didn't start with a flourish—he had trouble at Atlanta, Darlington, and Bristol. But he needed only two months to prove his worth. In the 12th

Photo: Harold Hinson Photography

start of his Winston Cup career, Earnhardt Jr. emerged victorious in the DirecTV 500 April 2 at Texas Motor Speedway, ironically the track where he had scored his first Busch win. For those interested in comparisons (and many were), Tony Stewart needed 25 races to get his first win, and the late Davey Allison, one of the best young drivers to appear in Winston Cup racing, ran 13 before his first checkered flag. And Dale Sr.? His first win came in his 16th race.

None of those numbers mattered to Dale Earnhardt Jr. on that magical Sunday in the heart of Texas, however. Those listening on his radio frequency heard a loud "Wooooo!" from Earnhardt Jr. after he crossed the finish line. That was only the beginning of an over-the-top victory celebration.

Junior arrived in victory lane, joining his ecstatic crew. Rushing into the wild scene was Earnhardt Sr., who pushed his way through the crowd to stick his head into the car window and hug his son. He hadn't

Dale Jr.'s first victory at Texas Motor Speedway in 2000. *Photo: Harold Hinson Photography*

overloaded Junior with advice over the years, but now he had some: Enjoy the moment. There'll never be another like this one.

"He just told me he loved me and he wanted to make sure I took the time to enjoy this," Junior said later.

He climbed from the car, stood on the roof and yelled at the crowd, then jumped down and into the arms of Senior, who had finished seventh in the race. As a car owner, though, he was first.

Junior's paycheck read $374,675. He had joined a select group of people—those who have won a Winston Cup race. As he pointed out in interviews later, many drivers race forever and never reach that point. To win in Winston Cup—even once—is an achievement not to be taken lightly.

From there, the first season was a mix of celebration and sadness—both for Earnhardt Jr. and his team and NASCAR in general.

The year saw the sport rocked by the deaths of two young drivers—Adam Petty and Kenny Irwin Jr., both killed in crashes during practice at New Hampshire International Speedway. Reverberations from their deaths would follow for years to come. The news about Petty hit Junior particularly hard because they had been close friends and had shared the joys and stresses of being in famous racing families.

In May, Junior added a win at Richmond International Raceway, one of the sport's classic short tracks, sparking another exuberant celebration in victory lane. This one was extra special because Junior's mother, Brenda Jackson, was on the scene, one of the rare occasions when she was able to see her son race. She had driven to the track from her home in Chesapeake, Virginia.

That would be Earnhardt Jr.'s last official Winston Cup win of his first season, but there was more—much more—excitement ahead. It came on a hot Saturday

night at Lowe's (formerly Charlotte) Motor Speedway, where Junior had made his first start a year earlier.

The Texas victory qualified Junior and his team for the Winston, NASCAR's version of an all-star event. The speedway annually turns the race into an extravaganza, with fireworks, flying confetti, an overload of hoopla, and loud, pounding music. Additionally, the race is a "special event," meaning no Winston Cup points are in play. That loosens the leash a bit on drivers who are worried about conserving their cars for points. Just the environment for a guy like Dale Jr.

At evening's end, it was Earnhardt Jr. finishing in front, and the reward was huge—$515,000. The satisfaction was even greater, for he had shown patience and power under one of his sport's biggest

because the speedway's winner's circle had been closed for the week in honor of the late Adam Petty. That put Earnhardt Jr. only a few yards from his adoring fans, and they hooted and hollered long into the night as their hero and his teammates sprayed beer and

> "HE'S ACTUALLY MORE CALCULATED THAN I THOUGHT HE'D BE. HE'S AGGRESSIVE WHEN IT'S TIME TO BE AGGRESSIVE." —KEN SCHRADER

spotlights and in front of one of its biggest crowds. He became the first rookie driver to win the Winston—no small feat.

Junior picked off pretenders over the race's closing laps and then ran down leader Dale Jarrett with two laps to go to win the race easily. The victory lane celebration was held on the track's frontstretch

champagne in one of the zaniest victory celebrations of the season.

In August, the Earnhardt clan enjoyed another special day. Kerry, Dale's older half-brother, started the Winston Cup race at Michigan International Speedway along with Junior and their father. It was Kerry's first Winston Cup start and marked only the second time a father had participated in a Cup event with two sons.

The closing months of the season were tough ones for Junior and his team. The strengths of the spring disappeared, and Earnhardt Jr. finished the year in 16th place in points, meeting his goal of the top 20 but failing to capitalize on the momentum he had built in the middle part of the year. He also lost the Rookie of the Year battle to Kenseth. The burdens and pressures of the year eventually took their toll, and there were more than a few disagreements within the team.

"We all had ego problems," Earnhardt Jr. said later. "We all had personality problems. We kind of all lost respect for each other. We kind of let the success we had at the first of the year get to our heads. Then we couldn't repeat that like we wanted to. We never really pointed fingers at each other, but we let each other know we weren't happy. We need to grow up and be able to handle stuff like that in stride.

"We just ruined it for ourselves, really. We put too

CHASE FOR THE NASCAR SPRINT CUP

Photo: David Griffin Photography

much pressure on ourselves, and when we got to the

big dance we just made fools of ourselves, I think. I'd

race every week of the year if I was running good, but

when you don't run good and you have bad finishes,

those are the ones that make you want the year to be

over with."

After a short offseason, it would be time to return

to Daytona.

Junior is introduced to the crowd before the Winston in 2002. Eventual winner Ryan Newman edged Dale Jr. for the win. *Photo: Harold Hinson Photography*

CHAPTER 6

One POPULAR DUDE

"I told him I really think [he's] the modern day Steve McQueen, that great, sort of mysterious American hero."

—Sheryl Crow

Photo: Harold Hinson Photography

The girl's name was Lisa. "Just Lisa," she said.

She wore a light blue T-shirt and tight white shorts, contributing to her ability to stand out from the rest of the crowd waiting at the Bristol Motor Speedway infield crossover. The T-shirt offered this spray-painted message on the front: "Dale Jr. Marry Me."

At every track on the Winston Cup circuit, at every speedway crossover, at every garage gate, at every possible spot where their heroes might appear, fans congregate, looking, wishing, hoping for the chance to see, touch, get an autograph from—marry?—their favorite.

The crowd waiting for Dale Earnhardt Jr. is usually bigger, rowdier and more dedicated than most. He and his public relations people must carefully plot his movements in and around garage areas because the throng that travels in his wake is often so big and boisterous that it creates safety concerns—both for the driver and those around him. As he briskly strides across the garage from point A to point B, the throbbing

Photo: David Griffin Photography

knot of people moves with him as photos, caps, programs and other prized possessions are thrust in his face for the ultimate validation—the Dale Jr. autograph. He walks along and signs, a practice perfected by all leading drivers.

Few drivers attracted the massive fan following Earnhardt Jr. could claim at such a young age. At most Winston Cup tracks, the cheers that follow him across the stage at driver introductions are louder and longer than those for virtually any other driver. At autograph appearances at malls, automobile shows, and similar functions, he faces lines that snake far into the distance, lines built by fans who will stand for hours—some arriving very, very early to get good spots—for a moment with their hero.

How did the Dale Jr. phenomenon happen so quickly? There are three answers.

He's good.

He's a cool, happening, in-touch dude.

Fans mob Dale Jr.'s souvenir trailer at Lowe's Motor Speedway. *Photo: Harold Hinson Photography*

Last, and certainly not least, he's an Earnhardt.

When Junior appeared in Busch Grand National racing and quickly showed promise, he automatically gained the allegiance of thousands of fans who had followed his father with fierce devotion for many years. It certainly didn't hurt that Junior drove cars

If Dale Sr. represented the Baby Boom generation and the wave of blue-collar fans who wanted a hero who came from their ranks, Junior is Generation X, Y and Z. He likes off-the-wall rock and rap music, dresses in baggy jeans and obscure T-shirts, often wears his caps backward and even has been known to dye his hair

> "HE'S EXTRAORDINARY. I NEVER COUNT HIM OUT. A LOT OF THINGS HE DOES REMIND ME ABOUT HIS DAD."
> —JEFF GORDON

owned and fielded by his dad's Dale Earnhardt Inc. NASCAR operation.

But Junior's popularity stretches beyond family ties. He is very much his own man, in many ways like his father but in others strikingly different. And that, too, has contributed to his fan appeal.

on occasion (a thought that would never have entered the mind of his father). Dale Sr. liked hunting; Dale Jr. likes computers.

This Earnhardt is Dale Junior, Official Cool Dude. A party guy with fast cars.

"He is just so cute," said Lisa, the girl with no last

name. "He drives fast. He wins races. He looks so good in that uniform. How could you pull for anybody else?"

Indeed, a quick trip through the souvenir sales area at any Winston Cup track might make one wonder if anybody else is being pulled for. The crowds at Dale Jr.'s souvenir trailers—where T-shirts go for $25 and $30, caps for $20, jackets for $150—are swarming hordes compared to those at most other driver sales spots. He and Jeff Gordon are the leading souvenir dollar producers in the sport.

In January 2002, at a benefit auction, two women paid $2,800 each for life-size cardboard standups of Earnhardt Jr. The purchases came with a kiss from The Man, which, many concluded, boosted the price considerably.

"I've been told I'm a pretty good kisser," Earnhardt said in a press conference later that day. "I guess there's a few more witnesses to that now, huh?" Then he grinned at the media gathering and said, "I don't see anybody paying $2,800 for you guys."

Part of the Earnhardt phenomenon is the fact that Junior is much more than a good race car driver. His reach goes far beyond the traditional NASCAR circle.

In 2002, Earnhardt Jr. was named one of *People Magazine's* "Most Eligible Bachelors." Remarkably, it was his sixth appearance in the publication. In 2000, he was part of *People's* "Sexiest Men" issue, and in 2001, he was judged one of the year's "Most Intriguing People."

In 2002, *Sports Illustrated for Women* conducted a poll asking fans which male athlete they would prefer to see in a swimsuit, a not-so-subtle twist on the weekly *Sports Illustrated Magazine's* annual display of the female figure in its famous swimsuit issue. Earnhardt Jr. finished seventh on the list.

Dale Jr. has been a subject of MTV's *True Life* documentary series. Television stars like Pamela Anderson, music biggies like Kid Rock and Edwin McCain, and other celebrities inhabit his pit at big races.

Photo: Chris Green

In July 2002, Earnhardt Jr. received one of sports' most visible tributes, appearing on the cover of *Sports Illustrated* as the featured figure in a long article about NASCAR's continuing growth.

Driver No. 8, the book he coauthored with Jade Gurss, his public relations guru, about his rookie Winston Cup season, roared to the top levels of *The New York Times* Best Sellers list and kept Junior busy signing its front pages.

He has appeared in music videos with Sheryl

Dale Earnhardt Jr. and the rock band Smash Mouth at The Rock in 2001. *Photo: Harold Hinson Photography*

Crow and the Matthew Good Band. He hangs with the in crowd practically everywhere he goes. In December 2001, he was chosen to carry the Olympic torch through part of Charlotte, N.C.

In addition to Anheuser-Busch (his primary sponsor), other companies, including some outside the traditional NASCAR mainstream, have joined the Earnhardt rush. Drakkar Noir cologne, for example, uses Junior's super-cool, black-leather-jacket image in its magazine advertisements, ones that urge consumers to "feel the power." His deal with Drakkar Noir includes personal appearances where he signs bottles of the cologne for hundreds of people who line up for the chance to share a word or two with him.

"The bottom line to Earnhardt Jr.'s appeal is that he has so many things going for him," said NASCAR vice president Jim Hunter, who saw Junior's grandfather and father race to the top of their sport.

"The first reason he has had a big impact is that

Pamela Anderson wishes Junior good luck before a race in 2002. *Photo: Harold Hinson Photography*

he is Dale Earnhardt's son," Hunter said. "Coupled with that, he has been successful. And he's extremely intelligent. He has street smarts beyond his years. He's matured in some ways, and he hasn't in others, and I think that's one of the things that makes him so appealing to young people.

"Young people can really identify with Dale. He's insightful and very mature in a lot of ways, but when it comes to drinking beer and being one of the boys, he's a 20-whatever-year-old guy."

Earnhardt Jr. established his party persona early on. He converted the basement level of his Mooresville, N.C. home into a bar and music club, a spot that soon became known as Club E. Parties there stretched deep into the night and eventually reached a level of intensity that moved Dale Jr. to trim their size and exuberance. A notorious late sleeper with a busy racing and appearance schedule, he also found he couldn't continue to burn the candle at both ends.

Still, he has retained the image of a freewheeling, single, good ol' boy with hard-rock music pounding in his head and a case of Bud forever frigid in the cooler.

"He has that Huckleberry Finn sort of realism that tells you that what you see is what you get," Hunter said. "He tries to be extremely honest. That sometimes works for him and sometimes works against him. To younger people, that's probably refreshing because they identify with that."

Hunter said Junior compares favorably to Senior at the same point in their careers. "Dale did many of the same things that Dale Jr. has been criticized for doing, and that's no different than any other father-son that I've known of," Hunter said. "It's life. I think his insight is a trait that his dad had. He 'gets' it. When everybody wants a piece of you, you have to understand that. If you weren't driving a race car, they wouldn't want a piece of you, and he gets that."

There is little doubt that Earnhardt Sr. was the most revered driver of his generation. After Richard Petty and before Jeff Gordon, he was the king of the show, the big man in town, and the predator who ruled all he surveyed. His driving style, his long run of success and his down-home North Carolina mannerisms made him a hero to millions. He was a blue-collar idol, an Elvis on wheels.

Over the years, it became possible to buy almost anything with the Dale Earnhardt name attached to it. The Dale Earnhardt leather sofa, for example (black, of

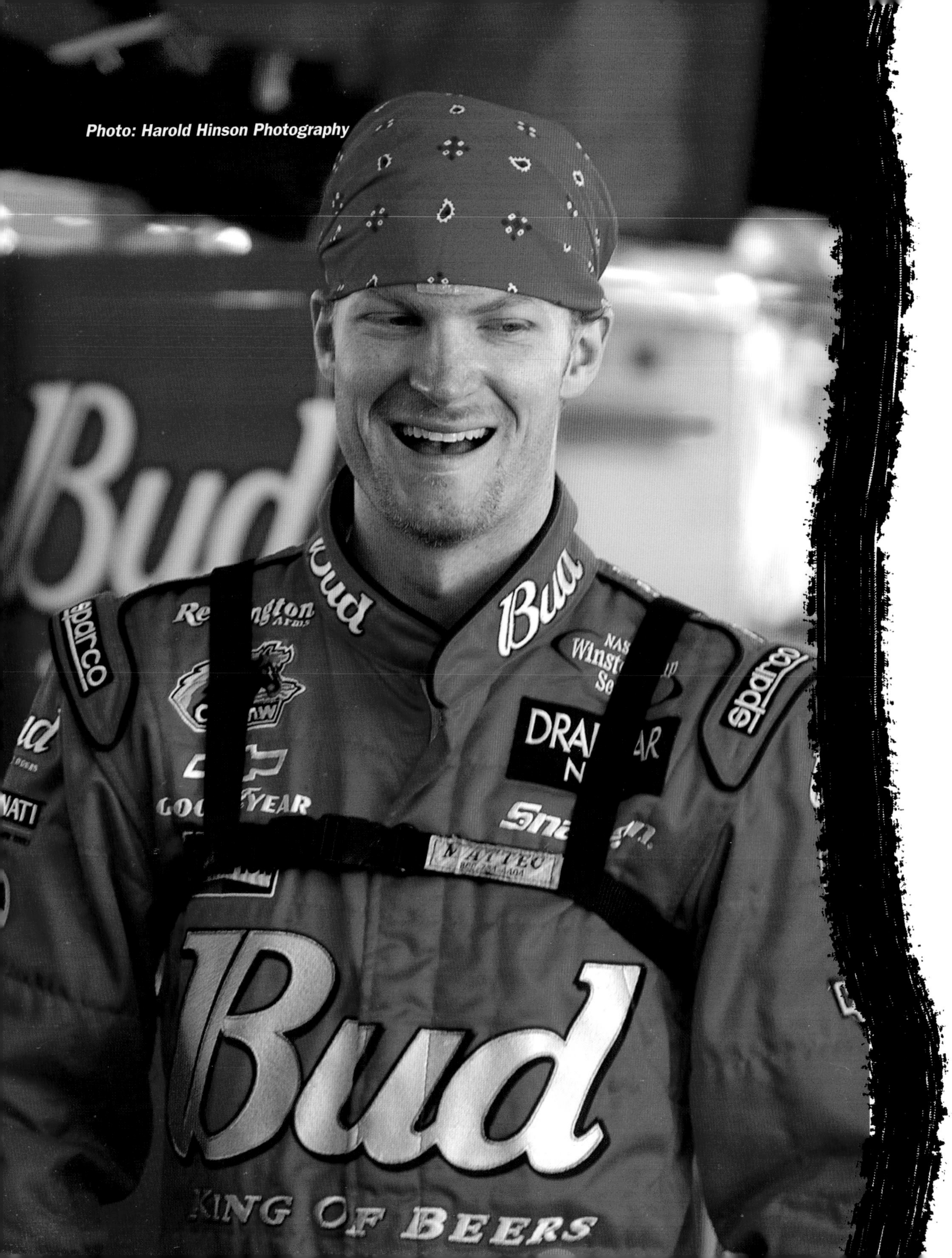

Photo: Harold Hinson Photography

course, with a big No. 3 in the middle of the back). Snap-on Tools produced a Dale Earnhardt "Intimidator" limited edition toolbox—with artwork by racing artist Sam Bass—similar to those used by NASCAR teams in the pits. More than four thousand were sold at $10,000 each. One devoted Earnhardt couple purchased a toolbox, made it an Earnhardt shrine of sorts in their modest home and saw it repossessed after only three payments. The wife cried as it was hauled out the door.

At Dale Earnhardt Chevrolet, the dealership Earnhardt owned in Newton, N.C., there was a waiting list to buy the used Silverado trucks Earnhardt would drive for a few thousand miles and return to the shop. To drive home in a truck the Intimidator had driven— that was something.

This is real devotion, and it was evident—still is, to a large degree—everywhere NASCAR races.

Although not every Earnhardt Sr. fan jumped on the Junior bandwagon, many did. A familiar scene in

Photo: Harold Hinson Photography

the traffic lines entering any Winston Cup race is a car with a 3 flag on one side and an 8 on the other, perhaps matched with an "Earnhardt Forever" bumper sticker.

"I'm going to have some fans that are going to root for me because I'm my father's son," Junior said.

"That's cool. I appreciate that. It makes it easier for me. You'll see them come in and they'll have Dale Earnhardt Jr. shirts on instead of Dale Earnhardt shirts. They're probably Earnhardt fans first and foremost, but they got to rooting for me and enjoying it."

Despite his popularity, however, Earnhardt Jr. clings to his Mooresville roots. Some of the best times of his life, he says, come with the small group of Mooresville friends he calls the Dirty Mo (short for Mooresville) Posse. They show up at races here and there and occasionally take wild and wacky road trips when Junior's schedule permits.

In the offseason between 2001 and 2002, Earnhardt and several friends "borrowed a truck" (as he put it) from the Earnhardt dealership and drove to Buffalo, N.Y., to move another friend to North Carolina. They left on the day after Christmas with no specific itinerary in mind—just a bunch of guys on a road trip.

"We got just past Greensboro [North Carolina] and turned off the interstate and just followed the compass from there on and never got on another four-lane road," Dale Jr. said. "It was pretty cool going through all those back roads. We went to Washington, D.C. and took our picture in front of the White House and several of the monuments. We went through Gettysburg and saw the battlefields, but it was about two in the morning, so it was kind of hard to see what was going on.

"It took us about 22 hours to get up there because we were fiddling around all day and all night. On the way back, we went through Ohio. We stopped at a bar and spent the night at a hotel and had a good time. It was a lot of fun. It was kind of cool to get away and be normal for a while."

Junior's youth and his approach to his job have earned him national publicity in places that traditionally have shunned stock car racing—Rolling Stone Magazine, for example, and MTV. He also sat for an interview with *Playboy* magazine, and he has explained his racing point of view on Jay Leno's late-night television show.

"I might not represent the average mold for a NASCAR driver, but we've been able to go to certain areas and certain people throughout the country and show ourselves and the sport to some interesting

sponsor for not having their logo all over my back and my shoulders and my head. But just walking around in a pair of Adidas is pretty cool to me, and I want everybody to know that's who I am. If you don't mind a guy that represents the sport wearing Adidas and a hat backward, that's fine. If you do mind, look

> "I'VE SPENT TIME WITH A COUPLE ROCK STARS AND I'M NOT AT THAT LEVEL. BUT I CAN IDENTIFY WITH WHAT THEY GO THROUGH A LITTLE BIT."
> —DALE EARNHARDT JR.

groups of people," Earnhardt Jr. said. He said he plans to continue to be himself despite the strategies of corporate America.

"I like to wear just normal clothes every day and be myself," he said. "I get a lot of flak from my

somewhere else.

"I don't think you can fool the public. When you dress somebody up and they do the dance, I think the public can tell the difference between somebody who's sincere and somebody who's not. I'm very proud of my relationship with all my sponsors, whether it be Budweiser or whoever, but I don't feel like it's necessary for me to don their logos everywhere I go. For some reason, I think we can go further just being ourselves, and I think people will be more interested in that than billboards."

Earnhardt Jr.'s "let-it-be" attitude spreads to his team, known as one of the most fun-loving groups in the sport.

"We're always so busy and so serious in this industry," said Jeff Clark, a veteran mechanic and the team's gasman. "A lot of times the fun gets left out. When you can do things like wear a hat backward like Dale does or wear the do-rags (as the team has done)

or a bandanna, it's all just to put some fun back in the game."

Part of the result, in addition to winning the occasional race, is that Earnhardt Jr. is helping to carry NASCAR forward with a new boatload of fans.

"I don't know if you'd call it a rock-and-roll crowd or what," Clark said, "but it's a younger crowd. He's branched out and built bridges to crowds that aren't normally NASCAR people. These people all will eventually have families. As he gets older, so will his interests and those of his fans. It's a good thing for the sport."

Photo: David Griffin Photography

Photo: Harold Hinson Photography

CHAPTER 7

TRAGEDY And REDEMPTION

"He was with me tonight. I don't know how I did it, but he was there. I'll be crying sooner or later — I don't know — I feel pretty good right now. I'm wore out now. I dedicate this win to him. I mean, there ain't nobody else I could dedicate it to that would mean more to me. I want to thank [his father's widow] Teresa back home — I hope she's loving this 'cause we sure are."
—Dale Earnhardt Jr. after winning the Pepsi 400 in NASCAR's first return to Daytona after Dale Sr.'s death.

WINSTON CUP CHAMPION
1980 · 1986 · 1987
1990 · 1991 · 1993 · 1994

Photo: Harold Hinson Photography

"HE'S CRAZY. WE'RE FROM TWO DIFFERENT GENERATIONS, BUT WE SEEM TO BE ABLE TO MESH PRETTY GOOD." —MICHAEL WALTRIP

Nothing could have prepared Dale Earnhardt Jr. for the crucible he would face in the 2001 season.

The year began with high hopes. In January, as DEI worked overtime to prepare cars for the first race of the new year, Earnhardt Jr. talked optimistically about the Daytona 500. He was confident he was going to win the race for the first time. The reason? He had had a rather vivid dream about Daytona, and he had finished first.

"Call me crazy, but I'll be talking to you at the postrace interview about how I did it," he said. "It was so real it was crazy."

The Daytona 500 was no dream, though. For Junior, the rest of the Earnhardt family, and the extended NASCAR community, it was a nightmare that was all too real.

With one lap to go in NASCAR's biggest race, everything appeared to be golden for Dale Earnhardt Inc. Two of the team drivers—Michael Waltrip and Junior—ran one-two, and Earnhardt Sr., driving, as always, for Richard Childress Racing, was in a huge pack of cars immediately behind them.

The race came down to Waltrip and Junior, and Waltrip, finally reaching the end of a frustratingly long winless streak, reached the finish line two car lengths in front. Behind them, though, Waltrip's boss and Junior's father had found trouble—big trouble.

Photo: Harold Hinson Photography

In a gaggle of cars trying to fight through the final turn on the final lap, Earnhardt had lost control of his car, been bashed in the passenger side by Ken Schrader and slammed headfirst into the outside wall. His familiar black Chevrolet fell down the banking and came to rest on the grass flanking the track apron.

Earnhardt was slumped in the car. Rescue workers arrived quickly and determined that the situation was grave. After the car's roof was torn away and Earnhardt was removed, he was transported to nearby Halifax Medical Center, where he was pronounced dead at 5:16 PM.

It was the hardest of ironies. Although dozens of drivers had been killed at Daytona over the years, no one had ever lost his life in the Daytona 500. Now Earnhardt, who had finally won his sport's biggest race in 1998 after two decades of trying, was the first victim. And at almost the same instant as his driver (Waltrip) and son crossed the finish line in a one-two flourish.

After the race, Junior was hustled from his car on

Michael Waltrip, right, shakes hands with teammate Dale Earnhardt Jr., center, as teammate Steve Park, left, looks on at the North Carolina Speedway in 2001.
Photo: AP Photo/Chuck Burton

Photo: Cpl. Stacey Bullock

pit road and escorted to the hospital. But he never saw his father alive again.

Only two weeks earlier, he had enjoyed one of his biggest days with Senior, sharing a ride with him in a Chevrolet Corvette in the 24 Hours of Daytona endurance race.

After the Black Sunday at Daytona, the sport wilted into mourning. One of its icons, perhaps its greatest driver ever, was gone. Junior had lost much more—a father, a hero, and a guide.

The next week was a dizzying mix of emotion. The three DEI teams returned to Mooresville, and all of the team's employees met at the shop Monday morning to hear team officials tell them they would race on, just as the boss would have wanted, they said. NASCAR had announced in a Monday press conference that the following weekend's schedule at the circuit's next stop, North Carolina Speedway in Rockingham, would proceed as planned. All three DEI drivers—Junior,

Waltrip and Steve Park—would participate, and Kevin Harvick was named to replace Earnhardt Sr. in the Richard Childress Racing car.

Dale Earnhardt Sr. was laid to rest on Wednesday in a private family ceremony, and a memorial service for family members, DEI employees, NASCAR drivers and officials, and others with ties to the sport was held on a rainy Thursday in Charlotte. The next day, everyone trudged to Rockingham.

On Friday at the speedway, Junior spoke to the public for the first time about his father's death. "I just want to thank all the fans and all the NASCAR family and everybody involved for the respect they've shown over the past several days," he said. "It's been a tough time. There have been a lot of questions and things running around in our minds. Our main focus is to try and maintain and progress with the vision my father had for Dale Earnhardt Inc. to the best of our ability.

"One of the things this teaches us is how selfish

Photo: Harold Hinson Photography

you are about things like this. I miss my father. I've cried for him out of my own selfish pity. I try to maintain a good focus for the future and remember that he's in a good place, a place we all want to be."

At the speedway, there was practice and qualifying, but the sense of loss was achingly real. "It's like waking up in the middle of the night and the electricity is off—and you can't find the light," said team owner Eddie Wood, a longtime Earnhardt friend.

Perhaps fittingly, the race was rained out on Sunday. On Monday, everyone returned and they held a race, the first one without Dale Earnhardt Sr. in more than 20 years.

Junior, still carrying the heavy burden of the loss of his father, crashed hard on the first lap. Eerily, he hit the outside wall in much the same fashion as his father had at Daytona one race earlier. He wasn't seriously hurt, but he finished last.

Steve Park, Junior's DEI teammate, made the day memorable by winning the race and dedicating the victory to the team's fallen leader. The win was celebrated up and down pit road, where drivers and mechanics had honored Earnhardt's memory by wearing specially designed "3" caps.

For Junior, the relentless passing of the season became a day-to-day trial. He was under the gun to perform for his team while also playing a role in trying to glue everything together after the incomprehensible loss of his father.

"Big E's death was a setback," said Jeff Clark, a No. 8 crewman. "That took the focus off things for a while. But I think our driver—his focus was always there. It's amazing to me how he could put his personal stuff on the shelf, so to speak, and focus on what we had to do. I was very impressed with that."

DEI manager Ty Norris said Junior was a key player in keeping the organization flowing in the weeks after the Daytona tragedy. "At first, I heard people say

DEI teammates Steve Park, Dale Jr. and Michael Waltrip at North Carolina Speedway in February 2001. Park and Waltrip are wearing the hats that were specially designed to honor Dale Earnhardt's memory. *Photo: Harold Hinson Photography*

that we can't do it without him [Earnhardt Sr.]," Norris said. "And we can't do it the same without him, I'll tell you that. That's where Teresa and Dale Jr. stepped in and said, 'Well, we're going to do it and we're going to do it right.' Because of that, that kind of worked everybody up."

Months of controversy followed the Daytona accident as NASCAR launched a long and expensive investigation into Earnhardt's death and a second-by-second recreation of its particulars. With a string of driver deaths caused by skull fractures, NASCAR was under heat to fix the problem quickly.

Before NASCAR issued its final report on Earnhardt's death in August, the circuit faced an emotional return to Daytona International Speedway, the site of his fatal accident, for the July 7 Pepsi 400. Revisiting the scene of the February tragedy would be difficult for everyone, but particularly so for Junior, who would forever carry the memory of his father's final Daytona 500 in his mind.

Junior went to Daytona Beach a few days early in the company of close friends, relaxed on the beach and visited the speedway. It helped him deal with the swirl of emotions, he said.

On race night, Earnhardt, whose season had been good but not spectacular, was looking for his first win of the year. It would be a storybook tale—the son winning the first race at the sport's most famous track five months after his father's death in its fourth turn.

And it came true.

In a sort of repeat of the 500, Junior and Michael Waltrip roared toward the finish in first and second. This time, however, the positions were reversed. Earnhardt, who led much of the night, showed the way to the finish, and Waltrip followed closely in a tight draft, crossing the finish line in Junior's shadow.

The crowd erupted in spasms of joy, and Earnhardt roared around the track for a cool-down lap before spinning onto the grass that separates the racing surface from pit road. He jumped from his car and onto its driver-side door and celebrated in uninhibited fashion as cheers from the grandstand washed over

"A LOT OF THINGS CHANGED HIS SEASON. BUT HE'S A GREAT DRIVER. I THINK HE'LL BOUNCE BACK NEXT YEAR, FOR SURE. I SEE HIM AS ONE OF THE GUYS TO BEAT." —DALE JARRETT

Photo: Bo Nash

Photo: David Griffin Photography

him. Waltrip stopped his car near Earnhardt's and joined the mad celebration, hugging Earnhardt as they stood on the roof of the winning car.

As the champagne flowed, a unique tribute to Earnhardt Sr. was unveiled in the sky over the speedway's infield. Fireworks exploded over the track, creating clouds of smoke, and a huge spotlight splashed the number 3 onto the cloud cover.

Ironically, Junior's final pass for the lead — around Johnny Benson — came at virtually the same spot in turn four where his father had crashed in February.

"The very first lap we went around the racetrack felt different and felt kind of rough," Earnhardt said. "But after the first run [in practice], it was just like always. I didn't really think much about what happened here in February or anything else."

Earnhardt's car was so fast that he didn't have time to think about a lot. He led an astounding 116 of 160 laps, a performance so dominant that it led to questions in the days after the race. Some suggested that NASCAR had given Junior a few breaks for the race, hoping to present a dramatic Earnhardt victory for the crowd and national television audience.

Junior was livid at such talk, saying the team had simply done a tremendous job of preparing a car to race at Daytona, a track where huge drafting packs are the norm.

On one dramatic night in July 2001, an Earnhardt wrote a big exception to that rule, taking the race and making it his own.

Much like another Earnhardt before him.

Photo: Harold Hinson Photography

CHAPTER 8

The NEXT TURN —And BEYOND

"I know that it'll be many, many years, and if I'm lucky to even come close to the comparison of what my father's done. But each thing I do is a step in that direction to be known as one of those good drivers. I want to be one of those 50 drivers in that next book. I want to go down as maybe one of the best. Right now, we're on track."

—Dale Earnhardt Jr.

Photo: AP Photo/Terry Renna

The therapy for Dale Earnhardt Jr. after the tragic circumstances of February 2001 was found on the racetrack.

It began with the electric victory at Daytona International Speedway in July of that year, and there would be two other similarly emotional wins to follow.

The first arrived in red, white, and blue fashion Sept. 23 in the MBNA 400 at Dover International Speedway in Delaware. The race was the first Winston Cup event after the Sept. 11 terrorist attacks in New York City and Washington, D.C., and emotions were still running high on race day. The track was awash in United States flags (spectators were given flags as they entered the stands), and many cars in the field carried red, white, and blue touches.

After a patriotic pre-race show, the field settled in for 400 rugged miles on one of NASCAR's toughest tracks. There were the usual meetings with the wall — par for the course on the fast one-mile oval — and the race came down to Earnhardt, who led 188 laps, and Jerry Nadeau at the end. Despite a late-race caution that

Dover winner Dale Earnhardt Jr. in September, 2001. *Photo: HHP/Harold Hinson*

Photo: US National Guard

Photo: David Griffin Photography

bunched the field, Earnhardt sprinted home with the win.

He drove to his pit, picked up a large American flag and took a reverse victory lap around the speedway with Old Glory flying high. In the stands, most of the assembled 140,000 waved their flags and cheered. It was one of the most emotional moments of the season.

A month later, Junior made it to victory lane at Talladega Superspeedway, winning the EA Sports 500, the final race his father had conquered. Senior's win came a year earlier in one of the most dramatic charges of his career as he moved from 18th to first in the race's closing laps.

Once again showing that DEI had the proper aero package for success at NASCAR's restrictor plate tracks, Junior stayed at or near the front virtually all day and made some bold moves to reclaim the lead late in the race. His win gave the organization his father built victories in three of the four plate races that season.

Earnhardt and his DEI mates were practically the only garage area residents who left Talladega happy.

A monster last-lap wreck that occurred behind Junior involved 17 cars and sent Bobby Labonte's Pontiac rolling onto its roof. NASCAR officials listened to complaint after complaint from drivers after the race and promised to devote more time and study to solutions to the "big wreck" problem at Talladega and Daytona.

As for Earnhardt, his Talladega dominance sparked memories of the cool confidence with which his father scored win after win at the series' restrictor plate tracks. "His dad was good at restrictor plate racing, as well, so maybe it's something he taught him or it's a natural knack that he just picked up," said driver Robby Gordon.

The Talladega victory carried a $1 million bonus, thanks to the No Bull Five program sponsored by the R. J. Reynolds Tobacco Co. More importantly, it was another confidence boost for Earnhardt.

"Toward the end of the year, when we won the No Bull race at Talladega, that meant a lot to him," crewman Jeff Clark said. "Success heals a lot of pains. That started getting his mind off what had happened [his father's accident]. I think as a person he grew up and matured a lot as well."

Earnhardt ended 2001 with three victories and almost $6 million in race winnings. He finished eighth in points, gaining eight positions from the previous season, despite being mired in 26th six races into the schedule. It wasn't the dominating year Earnhardt wanted, but the fiery victories at Daytona, Dover and Talladega were big building blocks.

As the 2002 season started, Earnhardt addressed the issue of pursuing a championship, a topic he generally had avoided. He knows, however, that there always will be comparisons to his late father and that he must strive to be championship timber, too.

"There's more to me, I think, professionally than just magazine covers and kick-ass sponsors and fun times," he said. "I want to win championships, and I like winning races. We look forward to winning more, but there's a side from just winning the championship and all the celebration and the extras and whatnot that come along with that.

"There's something to be said about having that asterisk beside your name for the rest of your life and

so I need to win some championships."

Becoming more serious about his driving, Earnhardt said, didn't mean giving up his fun times, some of which have become more or less legendary on the circuit. He can do both, he said, but it's evident that he is devoting more and more time to racetrack ins and outs and to the serious study of week-to-week

> ## "I AM THE SON OF ONE OF THE GREATEST DRIVERS. I'M NOT QUITE ONE OF THE GREATEST DRIVERS YET."
> ## —DALE EARNHARDT JR.

the rest of the time in the books that says you were a champion some time in your life. I can see that I have a great opportunity to really take it to several levels and to be somebody that maybe is in the same sentence with several of the greats in the sport down the road,

driving success.

"All that's still growing," said Clark. "I think he has a difficult time being younger, and he doesn't know where his place in the company is yet. That will all come with time. He has definitely made some steps to be a

Photo: David Griffin Photography

part of meetings that are going on and to learn where the management is taking the company."

As the 2002 season progressed, Earnhardt's positioning at DEI showed some not-so-subtle changes. As the year opened, he talked about "being pretty happy just being a race car driver right now," but later in the season, he placed himself in discussions about the futures of DEI drivers Steve Park and Michael Waltrip (both of whom were rumored to be on the way out of the organization but later signed contracts to stay aboard). And, in July, he even talked about the possibility of driving for another team, a move that wouldn't be unprecedented for a large family racing operation but one that would be particularly surprising for the son of Dale Earnhardt to make.

Although his stepmother, Teresa, owns and runs the company, Junior said he was becoming more involved with discussing matters with employees and learning about the day-to-day activities at one of motorsports' biggest operations.

Meanwhile, Junior wrestled with inconsistency through the middle months of 2002, winning—again—at Talladega and moving to fifth in Winston Cup points but then crashing hard in the next race at California, suffering a concussion. A few weeks later, he had dropped to 16th in the point standings.

With two-thirds of the season in the books, Earnhardt admitted to a sense of frustration and also said his focus and timing had been impacted by the concussion, a fact he didn't talk about until months later.

"Every driver in the garage gets to that point some time in the season," he said. "But we just got into a slump. We were fifth in the points, won Talladega, looked like we were going to have a shot at putting ourselves up there all year. We had been in the top five, the top 10, then had that real bad crash, real hard hit at California, and it took me as a driver about two or three weeks to get out from under that.

Dale Earnhardt Jr. at Charlotte Motor Speedway in Concord, NC in May 2002 Photo: HHP/Harold Hinson

"I'm not pointing fingers. That doesn't do us any favors. Didn't last year; didn't the year before. You get to a point where you run seven, eight races without a good finish, not a good thing happening to you, you get to a point where you either get up and work harder or you give up. I ain't giving up."

Earnhardt's thought processes in the heat of competition were illustrated well during the 2002 season in—perhaps surprisingly—two races in which he didn't finish first but still received accolades.

In the Winston all-star race in Charlotte, Junior rode behind eventual winner Ryan Newman entering the last lap and was within inches of Newman's car. He easily could have tapped Newman, moved him over and out of the way, and won one of the season's richest races. Instead, he drove Newman cleanly and settled for a tough second. Many of the assembled Earnhardt faithful in the crowd probably expected the bump-and-run move, but others said they gained a new respect for him for finishing the race without bullying Newman.

Earnhardt struggled with the decision after the race. "I'll play it over a thousand times in my head for weeks wondering how I could have done something different," he said. "I got a run on him and backed off. I should have stayed in the gas, and maybe I could have gotten inside him. Seven hundred and fifty thousand dollars is a lot of money, but it ain't worth all the people you piss off to get it."

In July at Daytona, there was yet another restrictor plate show for Michael Waltrip and Earnhardt, drivers who had dominated three of the past four races at the track. As the race neared its end, Waltrip and Earnhardt led a pack of drafting cars. If Earnhardt stayed put, he probably had second place wrapped up. If he pulled out of the draft to try to pass Waltrip, he might lose several positions.

In a move seldom seen in those days of jealously protecting one's position at Daytona, Earnhardt pulled out of the draft with three laps to go, hoping to challenge Waltrip with drafting help from behind. But no one chose to go with Earnhardt, and he finished sixth while Waltrip took the top prize home for DEI. Although it wasn't a win, Earnhardt scored some points with more than a few fans for at least giving it a shot.

"I stayed behind him [Waltrip] most of the race, kept us both up front, and when it came down to the last few laps, you've got to race," Earnhardt said. "I don't think anybody, when they really sit down and think about it, would want us to kind of just sit there and cookie-cutter the whole thing out and orchestrate the finish. I tried to make a move on him and almost had him, but Michael got more help."

Although Junior lost that gamble, just the fact that he was in position to make it is one of the things that attracts him to racing. Considering the name he carries, could it really have been any other way?

"I get such a thrill from it, and I enjoy it so much,"

he said. "It's well worth it to me, well worth the risk of losing your life or suffering some kind of serious injury. You never think about that when you're driving race cars. You never think about that when you're walking through the garage.

"You never think about that when you're sitting at home watching TV on the couch. It's definitely there. It's definitely a part of it. It's in your face sometimes when you don't want to look at it. But the fun and enjoyment we get out of it, it's worth it. There are a lot of things in life, a lot of things that people do that are dangerous, and I'm sure they have the same outlook. They love doing it, and they're willing to risk it.

"Plus, the pay isn't too bad."

So, Junior rolls on, always in the wake of his father, trying to establish a championship path of his own. Where will the road end?

"When will I retire? I don't know," he said. "I don't see myself going past 40. We'll see. Maybe I'll just choose to

take my motor coach and park it in the infield at Talladega,

turn up the stereo too loud and drink Bud with my friends

and just be a fan again. I can see myself doing that."

Dale Earnhardt Jr and Michael Waltrip at Phoenix International Raceway in Phoenix, AZ in 2003. *Photo: HHP/Harold Hinson*

156

CHAPTER 9

DRAMA AT DAYTONA

"For a moment, for a day, you're at the top of that mountain."
– Dale Earnhardt Jr.

Dale Jr. during driver intros to the Daytona 500 in Daytona Beach, FL in 2005. *Photo: David J. Griffin*

Some of Daytona's demons were banished for Dale Earnhardt Jr. in the summer of 2001, when he won the Coke Zero 400 at DIS. Only five months after his father's death at the track, Earnhardt Jr. was able to celebrate in the sport's most famous victory lane, and the tragedy of February was at least temporarily in the background.

But, to complete the circle at Daytona, to reach the heights his father had gained at stock car racing's cathedral, Earnhardt Jr. needed a win in the Daytona 500, the sport's classic race, its Super Bowl, its high ground.

That big moment arrived in 2004.

After February 2001, teams returned to Daytona each time with memories of Earnhardt still in the mix, and February 2004 was no different. But, particularly for the DEI teams, there was concentration on the task at hand, and Earnhardt Jr., thanks to his expertise on the circuit's big tracks and DEI's proven record in restrictor-plate racing, was considered among the race favorites.

He started third, ran up front most of the day, and then put himself in a comfortable position, drafting along with fellow Chevrolet driver Tony Stewart (a frequent Earnhardt drafting partner at the plate tracks) over the race's last segment. They swept across the finish line in that order, Earnhardt Jr. .273 of a second in front of Stewart.

The Daytona grandstands virtually exploded, the son having realized an accomplishment the father had scored in a moment of long-delayed glory only six years earlier. The line between the Earnhardts grew stronger.

The day's paycheck was $1.5 million, but that seemed almost secondary. The celebration went late into the night, but the full emotional impact of the win, Earnhardt Jr. said years later, took a while to hit.

"All the things that you want out of life, all the pressures that you put on yourself or you feel from other people, all the things you want to accomplish – everybody sort of has this mountain in front of them that they put in front of themselves they want to climb. … For a moment, for a day, you're at the top of that mountain," he told the media.

"All of your wants and all of your needs and problems you have, the little petty things that bother you, everything goes away. You just feel like you've realized your full potential. Everything is just sort of maxed out for the day, all the things you wanted to achieve.

"Obviously, you set a lot of goals for yourself, and that's just one of the goals, but just for a moment, just for that one day, whether it be 30 minutes or an hour after that race, as soon as you cross the finish line, you feel like it can't get any better than this. It's a pretty incredible emotion. I feel so lucky to have had that opportunity, experience. It is such a special, spe-

Dale Jr. during a press conference at the Michigan International Speedway in Brooklyn, MI. *Photo: David J. Griffin*

cial moment."

Members of Earnhardt Jr.'s team raced to the car to enjoy the moment with him near the track flagstand in the wild minutes after the checkered flag. It was one of those big sports moments of unbridled joy, everybody lost in a haze of happiness.

"Every time I see a replay of me and my crew

In winning the 500, Earnhardt Jr. was part of only the third father-son combination to score victories in NASCAR's biggest event. Lee and Richard Petty and Bobby and Davey Allison also had accomplished the feat.

The Daytona victory was the opener of a fine season for Earnhardt Jr. He won six races— a career

> ## "EVERYTHING IS JUST SORT OF MAXED OUT FOR THE DAY, ALL THE THINGS YOU WANTED TO ACHIEVE."
> ## — DALE EARNHARDT JR.

celebrating below the flagstand, it all comes back so clearly," Earnhardt Jr. said. "Every time I see it, I just think about how fortunate I was to have won that race. Some of the greatest drivers come through this sport and don't win it. It just doesn't seem right — only certain ones get to see that opportunity."

high — and finished fifth in points, backing up the third-place points run he had scored the previous season. He was the only driver to score top-10 finishes in each of the season's four restrictor-plate races, underlining again DEI's stature at Daytona and Talladega.

Included in the '04 surge was a win in the Oc-

Photo: David J. Griffin

tober race at Talladega Superspeedway, the Alabama track where the Earnhardt name is almost as popular as college football. Dale Earnhardt Sr. ruled the track during his peak years, and most of the Talladega fans who followed his every move transferred that allegiance to Junior.

He rewarded them in that fall race, leading 78 laps and outrunning Kevin Harvick by .117 of a sec-

ond to roar into victory lane.

It was a landmark victory for Earnhardt Jr., who remembered roaming the Talladega garage as a wide-eyed youngster on the heels of his dad.

"When I was a kid, I wasn't really concerned with how famous Dad was or the reaction from the fans," he told the media. "I was just running around the garage having fun with my buddies. I was in awe about how many brand new spark plugs they would leave laying around the garage. Brand new bottles of brake cleaner and glass cleaner just discarded. Because, when the race started, they didn't leave their crash carts. They didn't have crash carts back then. You had one tool box and you took it to the pits, so when the race was going on, the garage was completely empty of everything. There was nothing there. They had cleaned it up, put everything away and took everything they needed to the pits.

"What they would leave behind would be brand-new spark plugs and brand-new cans of this, that, and the other. All kinds of brand new stuff just laying around. Me and my buddies would box it up or carry it over to [independent driver] Jimmy Means' truck so they could take it home. Like they needed it. I don't even know if Jimmy cared, but we thought we were doing something cool. That was all we did; we did that, and begged people to let us wax their race cars

and whatever we could get ourselves into."

From that day to victory lane, Earnhardt Jr. had come a long way, and he was ending 2004 with a splash.

The seasons that followed, however, were feast and famine, as the ties that bound Earnhardt Jr. to his family's team began to show weak links. He won only one race in 2005, finishing a lowly 19th in points. He rebounded to fifth in points in 2006 but still won only once.

By 2007, it had become clear that Earnhardt Jr.'s future would be found outside the gates at DEI. He went winless that season in what became his final year at the team his father had started.

What once was thought unimaginable was about to happen. Dale Earnhardt Jr. would go racing from another address.

CHAPTER 10

A NEW DAY

"Aside from winning a few more races, I don't know how much more of a statement I could have made than what we made this year."
– Dale Earnhardt Jr.

Photography

Dale Earnhardt Jr. didn't have the opportunity to grow up in his own way and in his own time.

Almost from the day he first strapped into a race car, it was as if a bright, shining light hovered over his every move. He carried the Earnhardt name, of course, and that implied excellence. Everyone with any interest in auto racing wanted to know about his next move, his next win, his next plan.

There was little privacy involved.

"I can't mature as fast as I'm getting old," Earnhardt Jr. told the media. "[When] I was around the race track [as a kid], I really didn't have a whole lot of adult supervision. Once we got here, I could kind of scoot out from under Daddy's wing and do whatever we wanted— run around the track and goof off and have fun. I spent years and years doing that. It definitely is different—you might not be quite as mature as certain individuals in certain situations growing up around the race track."

But the pressures of big-time auto racing forced Earnhardt Jr. to address the Big Picture, to put serious thought into his place in the sport and to make tough decisions about the future.

The downslide at DEI and Junior's decision to depart the family team led to intense discussions about where he eventually would relocate, and the ultimate answer was not exactly a surprise. But the announcement was one of the biggest blockbusters in NASCAR history.

On June 13, 2007, Earnhardt Jr. announced that he had signed a five-year contract to drive for Hendrick Motorsports, at that time the sport's most productive team. There had been clear conflicts between Junior and his stepmother, Teresa Earnhardt, over the operation of DEI, and Junior said he felt his best shot at winning a Sprint Cup championship—his ultimate goal—would come elsewhere.

The break was not an easy one. Earnhardt Jr.

wanted to take his car number, 8, once driven by his grandfather, Ralph, to Hendrick Motorsports, but negotiations with Teresa Earnhardt over that move broke down, and Junior wound up driving No. 88. Sponsorships with Mountain Dew and the National Guard were announced, and Tony Eury Jr., Earnhardt Jr.'s cousin, made the move to Hendrick to remain Junior's crew chief.

The Earnhardt-Hendrick connection—a sort of gathering of giants in the sport—started in a rush of glory. In his first drive in the No. 88, Junior won the 2008 Budweiser Shootout, a non-points Daytona International Speedway event that traditionally begins the season. The next week, as excitement built toward the Daytona 500, Earnhardt won again in one of the Gatorade Duel races, which help to establish the starting grid for the 500.

The Earnhardt-to-Hendrick story would have continued in Hollywood fashion with a win in the 500, but Junior finished ninth in the season's most important race.

That set an unfortunate trend for Earnhardt Jr.'s first year with Hendrick. Although there were great expectations, the season was mostly mediocre. Junior won at Michigan in June to end a 76-race winless string, but that would be his only win of the season, and he finished a disappointing 12th in the point standings. Hendrick teammate Jimmie Johnson won the championship.

For those who thought 2008 was simply an aberration and that Earnhardt Jr. just needed time to adjust to a new landscape, the 2009 season was a big wake-up call. It was one of the toughest years of Earnhardt Jr.'s career.

Twelve races into the season, with Junior winless and 19th in points, Lance McGrew replaced Eury Jr. in the crew chief slot. But Earnhardt's struggles continued. He ultimately finished a stunning 25th—a career low—in points, missing the season-ending Chase for the Sprint Cup and recording only the second winless season of his full-time Cup career. The Hendrick buzz was fading.

In his first five seasons in Cup, all at DEI, Junior won 15 times and had an average points finish of 8.6. Over his final three seasons at DEI and his first two at Hendrick, he won three races and had an average points finish of 15.4.

Hendrick attacked the Earnhardt issues while keeping the rest of his organization fine-tuned. Johnson won another championship in 2009, and teammates Mark Martin and Jeff Gordon finished second and third in the standings, making Earnhardt Jr.'s positioning even more difficult to understand.

"We have some work to do there, but we're ready," Hendrick said. "To me, that's my challenge for 2010. We know we can be a little better in all three teams, but the real chore and the real job and the one we're going to tackle over there is making sure the 88 is where it needs to be."

The drought hit Earnhardt Jr. hard. He confessed at one point during the season to being "at the end of my rope."

The 2010 season brought little new light. Junior again went winless, much to the distress of his legions of fans. Although the year started with promise—a second-place run in the Daytona 500, Earnhardt failed to stay in the top 10 in points, missing the Chase.

Two days after the long season ended, Hendrick Motorsports announced yet another crew-chief change for the No. 88, with Hendrick veteran Steve Letarte named to manage Earnhardt's team for the 2011 season.

The new year again found Earnhardt Jr. frustrated in the victory column, but there were significant gains as he and Letarte built a strong level of communication. He came within a half-lap of ending what then was a 104-race winless streak in the Coca-Cola 600 at Charlotte Motor Speedway, running out of fuel with the lead and the checkered flag in sight.

Earnhardt finished seventh in points and gained a new high in consistency, finishing 99 percent of the laps run.

The arrival of Letarte made a big difference,

wouldn't expect anything less than him being a professional, as well. I think we have a good in-race relationship. He does a really good job of providing me with information and calming me that we are going to fix any issues we have.

"I feel confident that he has fixed enough issues and improved the car during enough races that

> "I KNOW WHAT I NEED TO BE FOCUSING ON, AND IT'S NOT WHETHER I GOT BACK AT SOMEBODY OR VINDICATED MYSELF AS MUCH AS JUST FOCUSING ON WHAT WE ARE TRYING TO DO TODAY, THIS WEEKEND, NEXT WEEKEND, AND SO FORTH."
> — DALE EARNHARDT JR.

Earnhardt Jr. said.

"I think I have gotten a lot better since working with Steve," he told the media. "He's definitely made me more accountable…. He's not going to put up with me verbally abusing him or the equipment. I

I don't really get as worried about it when something isn't quite right. I know that the chances of it getting improved and fixed are really good. I really don't even have to get in a conversation with him about it. He's pretty talented. He does a great job. He's really

Dale Jr. during driver introductions for the Sylvania 300 race at Loudon, NH. *Photo: David J. Griffin*

on top of what his responsibilities are. He manages his team really well. I've got great confidence in him and his abilities to orchestrate the weekend as good as I would expect. We get along really good because of that confidence between each other. I think there is good trust there, too."

Despite the gains, Earnhardt Jr.'s winless streak stretched into the middle months of 2012. Three days before the June 17 race at Michigan, he was asked about the season.

"I feel better right now than I have in the last several years when we weren't competing well, when we weren't running well and we had to answer as to why we weren't winning," he told the media. "We were miles from winning, you know? We were so far away from being able to compete and win a race and be competitive enough to win a race that that was a tough question to answer.

"Now, it just feels like it's right around the corner. So I'm getting excited. I'm getting more and more excited the more we run this year. The last couple of weeks we seem to have improved more as a team. Last year, we kind of held steady all year and we did well, and we were happy how we finished in the points, but we really just kind of held it steady the entire season. We didn't really find speed throughout the year. But this year, we've been able to start off good and now in the summer here in the last couple of weeks, we've found good speed more than we really had at the start of the year, which is really what teams that I see win races, do."

And, indeed, it was "right around the corner." Earnhardt Jr. brought Junior Nation to its feet by win-

ning the Quicken Loans 400 at Michigan, leading a solid 95 laps on the way to an easy victory over Tony Stewart. The seemingly endless winless streak finally ended at 143 races.

There had been talk that Earnhardt Jr. might never win again.

"It didn't really bother me," he told the media. "I was worried about my own situation more than anything, probably — what I needed to do to get it turned around. But I do feel a little bit vindicated to the people that considered I wouldn't ever be competitive again. Aside from winning a few more races, I don't know how much more of a statement I could have made than what we made this year. That's really not even secondary, though. That's not quite as important to me as just trying to make the best of this year with the final result being more wins and a championship. The closer we get to the Chase, the more real the opportunity seems. So, all that stuff comes to the forefront. I know what I need to be focusing on, and it's not whether I got back at somebody or vindicated myself as much as just focusing on what we are trying to do today, this weekend, next weekend, and so forth."

But, even as Earnhardt Jr. celebrated the win and a season of growth, trouble was on the horizon. During an August 29 test at Kansas Speedway, he hit the wall. Observers described it as one of the hardest crashes they had seen at the 1.5-mile track. But Junior left the speedway apparently OK.

Turned out that "apparently" was the active word. Soon, Earnhardt Jr. would be in the middle of a discussion that crossed lines between sports.

The Kansas incident was mostly forgotten until another vicious crash near the end of the Oct. 7 race at Talladega Superspeedway, a track infamous for its massive and dangerous accidents. Earnhardt Jr. was involved on the tail end of one of the biggest multi-car wrecks in recent NASCAR history. His car was tapped and spun as drivers tried to avoid the crashing vehicles in front of them.

Two days later, Earnhardt Jr. was experiencing headaches and a sense of uneasiness, apparently because of the Talladega crash. He made the tough decision—one most NASCAR drivers try every detour to avoid—to seek medical advice. The result? Tests confirmed a concussion, plus lingering effects from a concussion suffered in the testing crash at Kansas.

"I thought I was in the clear, but just that little accident at Talladega, I started having headaches and stuff immediately after the wreck, and then into the next day and into Tuesday, and I thought, man, this is pretty soon after the other accident in Kansas," Earnhardt Jr. told the media. "I should probably take this really seriously and seek some professional opinions on this."

The Earnhardt situation developed while National Football League officials and teams were in-

Dale Earnhardt Jr wins the Quicken Loans 400 at Michigan International Speedway in Brooklyn, MI in 2012.
Photo: HHP/Harold Hinson

volved in intense discussions about head injuries and concussions in football, a dialogue that spread across all sorts of media across the country.

Earnhardt Jr. sat out the next two weeks—at Kansas and Charlotte, in effect giving up any shot at success in the Chase and a championship.

The impact shot through the sport. NASCAR's most popular driver (a title he would win for the 10th straight time at the end of the season) was on the sidelines for two weeks. Some fans who had tickets for the upcoming races quickly made other entertainment choices, saying they had no incentive to be in the stands if Junior wasn't going to be on the track.

The fans weren't the only ones who were uncomfortable.

"I'm really going to feel pretty odd not being in the car," Earnhardt Jr. told the media. "I'm anxious, real, real anxious just to get back into the car and get back to—I think you learn not to take things for granted, and I just hate that this has caused such a fuss."

Earnhardt Jr. returned to the No. 88 cockpit for the Martinsville race and reported no after-effects from the Talladega crash. He finished the season more than ready, he said, for another strong start in 2013, and he dismissed previous thinking that he might retire from driving at 40.

The journey continues.